An Archaeology of Social Space

Analyzing Coffee Plantations in Jamaica's Blue Mountains

CONTRIBUTIONS TO GLOBAL HISTORICAL ARCHAEOLOGY

Series Editor:
Charles E. Orser, Jr., *Illinois State University, Normal, Illinois*

A HISTORICAL ARCHAEOLOGY OF THE MODERN WORLD
Charles E. Orser, Jr.

ARCHAEOLOGY AND THE CAPITALIST WORLD SYSTEM: A Study from Russian America
Aron L. Crowell

AN ARCHAEOLOGY OF SOCIAL SPACE: Analyzing Coffee Plantations in Jamaica's Blue Mountains
James A. Delle

BETWEEN ARTIFACTS AND TEXTS: Historical Archaeology in Global Perspective
Anders Andrén

CULTURE CHANGE AND THE NEW TECHNOLOGY: An Archaeology of the Early American Industrial Era
Paul A. Shackel

LANDSCAPE TRANSFORMATIONS AND THE ARCHAEOLOGY OF IMPACT: Social Disruption and State Formation in Southern Africa
Warren R. Perry

A Continuation Order Plan is available for this series. A continuation order will bring delivery of each new volume immediately upon publication. Volumes are billed only upon actual shipment. For further information please contact the publisher.

An Archaeology of Social Space

Analyzing Coffee Plantations in Jamaica's Blue Mountains

James A. Delle

New York University
New York, New York

SPRINGER SCIENCE+BUSINESS MEDIA, LLC

Library of Congress Cataloging-in-Publication Data

ISBN 978-1-4757-9161-7 ISBN 978-1-4757-9159-4 (eBook)
DOI 10.1007/978-1-4757-9159-4

Originally published by Plenum Press, New York in 1998
Softcover reprint of the hardcover 1st edition 1998

http://www.plenum.com

10 9 8 7 6 5 4 3 2 1

To the memory of my late father
George V. Delle

Foreword

James Delle has solved a number of problems in Caribbean archaeology with *An Archaeology of Social Space*. He deals with most of the problems by using historical archaeology, and clearly implicates Americanist prehistorians. Although this book is about coffee plantations in the Blue Mountains area of Jamaica, it is actually about the whole Caribbean. Just as it is about all archaeology, not only historical archaeology, it is also a book about colonialism and national independence and how these two enormous events happened in the context of eighteenth and nineteenth century capitalism.

The first issue raised appears to be an academic topic that has come to be known as landscape archaeology. Landscape archaeology considers the planned spaces around living places. The topic is big, comprehensive, and new within historical archaeology. Its fundamental insight is that in the early modern and modern worlds everything within view could be made into money. Seeing occurs in space and from 1450, or a little before, everything that could be seen could, potentially, be measured. The measuring—and the accompanying culture of recording called a scriptural economy—became a way of controlling people in space, for a profit.

Dr. Delle thus explores maps, local philosophies of settlement, town dwelling, housing, and the actual condition of plantations and their buildings now, so as to describe coffee-Jamaica from 1790–1860. He does a captivating job. He plots the changes, the ideals, the utter venality, and the supreme boredom of plantation space. Everything he describes was done to produce coffee beans—but also not to produce coffee beans. James Delle uses all the sources to reveal how the British and French nations took an island and treated it like a head of beef cattle. They measured it, parceled it out, cut it up, evaluated it, fattened it up with slave labor, took it to the market, slaughtered it, and then ate some and threw some out. They did it all rationally, and Dr. Delle records that. They did it all outrageously and on an enormous sale, and Delle records some of that too.

Beyond Jamaica and beyond coffee, Delle has begun to make landscape a way of doing historical archaeology that constitutes a voice against colonialism. He does this in many different ways. He understands that the key to power is a place in history and the key to history is the production of narratives. Delle has learned a lot from Michel-Rolph Trouillot, particularly from his *Silencing the Past*. As a result, Delle walked loudly—rather, definitely—into the silence of coffee production and asked about the slavery used to produce a cup of coffee. Delle shows that Caribbean slavery is a domain of scholarly silence, when it comes to coffee. Of course, what we all know is that historical archaeology is very active in understanding Caribbean plantation life and slave life.

This book examines the struggle for power through coffee – Jamaica between European planters/owners and the slaves of African descent who will not be slaves for long. Delle shows why and how coffee plantations were set up: How they were run, why they failed in about two generations, and what happened when they failed. He describes the desire of workers for self-sufficiency and how that easy-to-relate-to aim is debased—so reasonably—by saying that small independent owners have no capital and deprive the economy of a laboring base.

Perhaps the most moving part of Delle's book is the story of how plantation owners, once coffee failed, started to charge rent for the slave's and freemen's use of provisioning lands—the garden plots they had been allotted for growing their own food. They had to grow their own food because the owners took no responsibility for feeding them. Then the owners realized they owned the provisioning land, no matter how marginal it had originally been. Rents attacked the subsistence base of the workers, further alienated them, provided only temporary income for land owners, but exposed the owners for the exploiters they were.

Michele-Rolph Trouillot can be made to induce a historical archaeology that reverses the erasures in Caribbean history. He was thinking of the nonevent we have made of the slave revolt in Haiti in the late eighteenth and early nineteenth century that created Haitian independence. His point is that we cannot stand to know, let alone celebrate widely, the fact that, after ourselves, the second American nation to declare independence successfully was a large body of slaves in Haiti. James Delle takes a part of the process of emancipation and independence and moves it to Jamaica and traces it through historical archaeology. He traces liberation and simultaneously helps liberate historical archaeology from the shackles of the concrete and thus of banality.

Mark P. Leone
Department of Anthropology
University of Maryland, College Park

Preface

The Jamaican political economy experienced a series of structural crises between 1790 and 1865 that precipitated changes in the relations of production on the island. Faced with changes within the global circulation of capital, groups of Jamaican elites, using their positions of privilege within the socioeconomic hierarchy of the island, attempted to manipulate the socioeconomic upheavals of the nineteenth century to maintain and reinforce their wealth, power, and status within Jamaican society. Within this context, large-scale coffee production, first using slave- and then later waged-based labor systems, was introduced to Jamaica for the first time. This book analyzes how the introduction and development of this industry in one coffee producing region — the Yallahs drainage of the Blue Mountains in the southeastern quadrant of the island — affected the lives of people living and working in this upland region of Jamaica for the first time.

The introduction and development of this new industry forcefully altered the material and social spaces of the Yallahs region. This book considers two sequential phases of sociospatial negotiation: the introduction of coffee production under slavery and the reorganization of labor–capital relations following emancipation. The intentions behind, and the often contested results of, the elites' attempts at restructuring the logic of accumulation during these phases of manipulation are interpreted by examining the historical, cartographic, and archaeological records. These various data sets are considered to be manifestations of three interrelated dimensions of space: the cognitive, the social, and the material. By examining plantation space in this theoretical context, this study interprets the way new spaces were designed by elites to reinforce new social relations, and how such manipulations were resisted by the African-Jamaican majority of the Yallahs region.

Before delving into this analysis, I would like to briefly explain how and why the research presented here has taken the shape it has. When casual friends and acquaintances inquire about the specifics of my work, I very rarely get beyond the phrase "I am examining coffee plantations in Jamaica" before I detect signs of envy in the faces of a given audience; I am usually interrupted with a gleeful "Wow! That must be great!" or "How did you manage to get such a great project?" Such responses are spurred on, no doubt, by the visions of Jamaica as a land of swaying palms, sandy beaches, and reggae music, that have been created and perpetuated by the North American media. Such misleading images of Jamaica make it very difficult to explain the realities of life as experienced by most Jamaican people, especially considering that most North Americans who actually travel to Jamaica see little more of the island than the sanitized all-inclusive hotels, tourist markets, and tourist sections of Montego Bay, Negril, or Ocho Rios. Most of these places — including the beaches owned by tourist hotels — are off-limits to ordinary Jamaicans. Since most tourists never get to meet many ordinary Jamaicans, and tourist propaganda paints a rosy and peaceful picture of a tranquil island paradise, it comes as little surprise that when imagining Jamaica most people think of swaying palms and not the shantytowns, poverty, and economic oppression that are the reality for tens of thousands of Jamaicans. When working on the island, however, it is virtually impossible to ignore the harsh conditions under which so many people live. In order to understand the nature of this material and social inequality, it is necessary to understand the historical trajectory of the island.

Why, though, did I choose to analyze how global inequality developed on Jamaica? Part of the answer lies in my early exposure to Jamaican culture, and an early realization that much of the material prosperity enjoyed by North Americans is produced through the exploitation of peoples living in what is now commonly called "the developing world."

During my youth, the town in which I grew up still had a significant agricultural base; what are now suburban subdivisions were then farms and orchards. I spent several seasons of my youth working on an apple orchard: fertilizing and weeding trees, maintaining and placing apple bins, cleaning and operating a cider press, picking and sorting apples. It was while I was employed at this orchard that I made my first Jamaican acquaintances: migrant workers who came every August to pick apples, at a considerably lower wage than I was receiving. My hometown friends and I were always amazed at how quickly and diligently the Jamaican workers picked apples. It was common, at the end of the day, to realize that each Jamaican worker had picked three or four times the number of apples that I had picked. Even though I would be exhausted, the Jamaicans never seemed tired.

Working side by side with 10 or 15 Jamaican migrant workers gave me an early opportunity to develop my ethnographic skills. I was fascinated with the stories about migrant work and about Jamaica that these men would tell me in the fields; to my supervisor's vexation, I was much more interested in talking to these men than I was in picking apples! After work, my hometown friends and I would often hang out with our Jamaican co-workers, who were legally restricted from leaving the orchard at night. The orchard provided the migrant workers with what I now realize was fairly comfortable housing by migrant standards: a two-room apartment with a sink, stove, and refrigerator, in which as many as 12 or 15 men would live for the duration of the picking season. I well remember drinking our homemade hard cider (we called it "hooch") and store-bought beer, "reasoning" with my Jamaican friends, and learning about the conditions of the migrant camps in Florida, where these men picked citrus fruits and tomatoes and cut cane during various seasons of the year. I remember thinking even then that the restrictions on their movements, and the subminimum wages allowable by law, were unnecessarily oppressive.

In many ways, Michael, Beetle, and the other Jamaican "apple-pickers" planted the seeds of this research 20 years ago. It was through these friendships that I first learned about the realities of international labor exploitation. At the age of 14, I had become aware of a Jamaica that is not shown on TV, a land that people choose to leave to make $200 a week, if they are lucky. A land in which, I later learned, thousands of people stand in line for days in order to apply for a migrant visa. A land in which many people cannot afford what are considered the barest essentials in the United States. A land that is still coping, as we are in North America, with the aftermath of African slavery. It is the purpose of the following chapters to analyze how we have come to this point — how these conditions were historically created — by examining the dynamics of, and resistance to, past oppressions.

Acknowledgments

When one has finished a project of this magnitude, it is difficult indeed to know where to begin when expressing professional and personal gratitude for assistance. I owe a debt of intellectual gratitude to many people, including first and foremost those mentors who guided me through graduate school. Bob Paynter, Martin Wobst, and Ben Howatt served ably as my University of Massachusetts dissertation committee; I am also grateful to many faculty members of the department of anthropology at the University of Massachusetts including Oriol Pi-Sunyer, Helán Page, Art Keene, Dena Dincauze, Alan Swedlund, Ralph Faulkingham, and Brooke Thomas. I would especially like to acknowledge the late Sylvia Forman, who was in many ways my first mentor at the University of Massachusetts. I am also indebted to members of my MA committee from the College of William and Mary: Rita Wright, Camille Wells, Norm Barka, and Marley Brown. Doug Armstrong of Syracuse University first introduced me to archaeology in Jamaica; without his assistance this project would never have been completed. I would also like to thank those who have read and commented on various manifestations of the thoughts that have resulted in this dissertation, including Chuck Orser, Sidney Mintz, Stan Green, Tom Patterson, Chris Tilley, Mick Monk, Matthew Johnson, and Leland Ferguson. Paul Shackel generously gave of his time to read an entire draft of the manuscript and offered very useful advice on its improvement. My thanks are extended to all those friends and colleagues—too numerous to mention—who have been part of the intellectual process that has led me to this point; special thanks go to Paul Mullins, Ed Hood, Jim Garman, Blythe Roveland, Barry Brenton, Bob Paine, Mike Volmar, Mary Robison, Amy Gazin-Schwartz, Sue Hyatt, Ann Marie Mires, Nancy Muller, Angèle Smith, Ken Kelly, Matt Reeves, Jim Peckham, Greg Cook, Amy Rubinstein, and Norman Stolzoff.

Logistical support for this project was provided by a number of people and institutions. I would like to thank the excellent staffs of the Jamaica Archives, the National Library of Jamaica, the University of the West Indies at Mona, the Public Records Office of England, the Lambeth Palace Library, the Kent County Archives, the National

Library of Scotland, and the British Library. I would also like to thank the Jamaica National Heritage Trust for all the help and support they have provided me, particularly Dorrick Gray, Spider Walters, Ywonne Edwards (now with Colonial Williamsburg), and Roderick Ebanks. Lance "Dougie" Douglas and Ms. Jaffne provided me with luxurious hospitality at "Stony Gut" during the last field season; I am forever in their debt. A number of landowners were generous enough to let me access their properties during the course of my study. Thanks go to John Allgrove, William "Castro" Campbell, Dr. Charles Deichman, Lloyd Dixon, Keeble Munn, and the Jamaica Forestry Department. My thanks go also to the many people in Jamaica who helped with various aspects of my study, including Difference, Mr. Percy White, Brother Wolf, Miss Dottie, Tay, Steelhead, Zes, White, and Rock. A hardy thanks also goes to my field crews, especially Jim Dooley, Jackie Watson, Mary Ann Levine, Spider Walters, Dorrick Gray, Mike Volmar, Burt Osterweis and Tim Fulton.

A project like this, although admittedly and consciously low-tech, requires financial support. I would like to acknowledge the Graduate School of the University of Massachusetts, which provided me with a fellowship to begin the process of writing this monograph, the Department of Anthropology of the University of Massachusetts, which supported my original trip to London with a European Program Fellowship, and the Wenner-Gren Foundation; this project was funded, in part, by a predoctoral grant from the Wenner-Gren Foundation for Anthropological Research. This monograph was prepared for publication while I held an appointment as a visiting assistant professor of anthropology at New York University. I am indebted to the faculty and staff of New York University, particularly Rita Wright, Pam Crabtree, Randy White, Fred Myers, and Bambi Schieffelin; I was ably assisted in preparing the manuscript for publication by Susan Malin-Boyce and Mark Smith.

Finally, I would like to thank my family, particularly my mother, Marjorie Leyden Delle, and my late father, George V. Delle, for the support they have offered me through the years. My final words of gratitude go to Dr. Mary Ann Levine for her never-ending support of this and all that I dare to do.

Contents

Tables, Figures, and Abbreviations

Tables

Figures

Abbreviations

AP	Accounts Produce
CO	Colonial Office
PP	Parliamentary Papers
Mackeson mss.	Letters of John Mackeson
SR	Slave Returns
STA	St. Andrew Parish Plantation Map
STDVM	St. David Vestry Minutes
STT	St. Thomas Parish Plantation Map

An Archaeology of Social Space

Analyzing Coffee Plantations in Jamaica's Blue Mountains

Introduction | 1

INTRODUCTION

Over the course of the past two decades, a variety of scholars have sought to expand the traditional geographical and theoretical scopes of historical archaeology. This series, Contributions to Global Historical Archaeology, is one result of the increase in the number of research projects in historical archaeology being conducted beyond the traditional boundaries of the field. In the earliest years of the field, the majority of historical field workers focused their attention on the eastern seaboard of North America, using traditional and conservative field methodologies to analyze the material culture of the recent past. In the late 1970s, following in the footsteps of Jim Deetz, perhaps the most influential historical archaeologist of the last generation, historical archaeologists began to examine sites beyond the plantations of the Southeast and the colonial sites of the Northeast. Within a decade, the scope of North American historical archaeology had expanded to include numerous studies of, for example, mining sites and boom towns in Nevada (e.g., Hardesty, 1988; Purser, 1991), missions and later frontier sites in California and elsewhere on the West Coast (e.g., Adams, 1977; Costello, 1991; Farnsworth, 1987; Pastron and Hattori, 1990), and the Spanish borderlands in southeastern North America and the Caribbean (e.g., Deagan, 1983; Thomas, 1993).

By the early 1990s, historical archaeology had become a truly international field of study. Although there had been some interest in international historical archaeology prior to this time (e.g., Armstrong, 1982, 1985; Bonner, 1974; Cotter, 1970; Dethlefsen, 1982; Handler and Lange, 1978; Pulsipher and Goodwin, 1982), the past decade has witnessed a remarkable increase in the number of historical studies conducted beyond the shores of mainland North America. In many of these cases, historical archaeologists study either the material remains of European colonialists or the material remains of indigenous peoples who were in contact with such colonizers. Such sites can, of course, be found in nearly every corner of the globe. Historical archaeology is now commonly practiced on colonial sites in, for example, the Caribbean (e.g., Armstrong, 1990; Delle, 1989, 1994; Handler, 1996; Howson, 1995; Pulsipher, 1994, Reeves, 1997), West Africa (e.g., De Corse, 1991; Kelly, 1995, 1997), sub-saharan Africa (e.g., Falk, 1991; Hall, 1992; Perry, 1996; Schrire, 1992), Europe (e.g., Hood, 1996; Johnson, 1993, 1996; Mangan, 1994; Orser, 1996), and South America (Jamieson, 1996; Orser, 1992, 1996; Rice, 1994, 1996a–c; Rice and Smith, 1989).

Just as many social anthropologists now recognize that the globalization of capitalism has had an irreversible impact on the development of indigenous cultures around the world (Bodley, 1990), historical archaeologists commonly realize that such colonial processes are historically rooted and are by nature international in scope. Many historical archaeologists with international projects have thus had as either an explicit or implicit goal the extension of the theories and methods of archaeology to questions concerning the historical processes of European capitalist expansion across the globe. This book is the result of one such study. With this volume I hope to make a contribution to the growth of the field of historical archaeology as an archaeology of capitalism. In doing so I hope to demonstrate how the methodologies of archaeological landscape and spatial analysis can be used to address historical questions concerning the international expansion of European capitalism and how the material remains of the recent past can be used to understand the evolution of this historical political economy. More specifically, I intend to demonstrate how global political economic phenomena have historically impacted people in one small rural area, the Yallahs River drainage of the Jamaican Blue Mountains.

This book focuses on the spaces of colonial production on coffee plantations in the eastern highlands of Jamaica. The approach I take assumes that late eighteenth- and early nineteenth-century British colonial production was a sector within international capitalism, and thus operated under a division of labor through which the production, distribution, and consumption of commodities significantly contributed to the definition of race, gender, and class identities. It has already been well-defined in the wider body of social science literature that international capitalism operates within a significant and sometimes severe class-based hierarchy, which concomitantly produces definitions of gender and race that are used to reify the structures of inequality. Under the capitalist system, membership in the elite classes is restricted, i.e., while the majority of the population will belong to a working class, a minority elite class will control the means of production. One of the goals of the elite class will be to organize production in such a way as to maximize the accumulation of wealth, generally at the expense of specific sectors of the working classes.

Despite the historical success of capitalism as a strategy for social organization, capitalism should not be considered a static monolithic entity. It is, rather, a continuously changing set of complex social relationships that periodically experiences significant episodes of crisis and restructuring. The thesis that drives this study contends that during such periods of crisis within capitalism, elites accelerate their attempts to reorganize the spaces of production in order to reorganize the relations of production, thus to secure and maintain their positions of socioeconomic dominance.

Because such reorganization of space usually includes the remapping of the lives of nonelites without their consent, workers often resist attempts to restructure the spaces of their lives, particularly when they feel that such impositions will benefit the elites at their expense. Alternative spaces will be defined, proactively and in direct resistance to the spatial definitions imposed by the elites. To understand how these phenomena are negotiated, space must be understood as a material tool that can be manipulated to various purposes. This book thus considers space not as a series of neutral backdrops to human action, but rather as a set of active forces within the processes of historical social change. Because space is in part a material entity, using an active definition of space suggests how the theories and methods of historical archaeology are relevant to the study of social process. As ours is a field that interprets the material culture of the recent past and as various manifestations of space leave behind material residue, historical archaeology is uniquely situated to analyze the material, cognitive, and social elements of European colonial expansion through the analysis of historical spaces.

To analyze how the British used space as a tool of exploitation within colonial Jamaica, this book examines some of the material remains of what can be defined as a formative period of transition within the organization of European capitalism: the transition from mercantile capitalism (or simply mercantilism) to competitive capitalism. Mercantilism had been the primary organizing principle of British capitalism from the late sixteenth century through the first third of the nineteenth century. During the mercantile phase of capitalism, the power of the nation-state to levy tariffs and to control the flow of commodities was used to ensure the development of powerful commercial monopolies and oligarchies.

As a result, in part, of the development of industrial technologies, the organizing principles of mercantile capitalism were transformed beginning in the late eighteenth century. During this time the British government's role in directing the flow of commodities began to wane in the face of "free trade." As the government dropped price protections for certain commodities, private firms and corporations entered supposedly free and open competition to set wages and face the vagaries of commodities markets (hence "competitive capitalism"). Experiments in free trade began as the British shifted the focus of their colonial program away from the Caribbean, the Carolinas, the Chesapeake, and New England to newly created colonies in the Far East, the Indian Ocean, and mainland Latin America. In the old colonies of the West Indies, the imperial shift to competitive capitalism began to take a toll on the interests of the elites even prior to the abolition of slavery. The West Indian planters faced an extended period of crisis that resulted in part from the development of several contradictions within the organization of colonial commodity production. Among these contradictions was the ideological dialectic between free trade, i.e., unrestricted commodities markets, and slave labor, i.e., radically restricted labor markets. The resulting synthesis, competitive capitalism, eventually resulted in the abolition of the African slave trade, slavery per se, and market protections for West Indian commodities.

As Jamaica was obviously part of this British colonial empire, the first half of the nineteenth century witnessed a significant transition within the logic of capitalism as it was practiced in Jamaica. This included not only a shift in social relations, for example the abolition of slavery and the creation of a rural proletariat, but a concomitant, and perhaps even defining, reconfiguration of space and spatial control. The focus of this book is the analysis of how the transition to competitive capitalism was materially expressed in Jamaican space, and how that material space in turn helped to define and constitute social expression.

HISTORICAL ARCHAEOLOGIES OF CAPITALISM

As should by now be obvious to the reader, I agree with those scholars who argue that historical archaeology can perhaps best be defined as the archaeology of capitalism (e.g., Johnson, 1991, 1996; Leone, 1995; Orser, 1996; Paynter, 1988). In recent years there has been an increasingly influential movement to so define historical archaeology. The search for an archaeology of capitalism has its roots not only in Marxist thought, but in the wider intellectual history of the field. Marxians and non-Marxians agree that historical archaeology should focus on the material expressions of European expansion. This intellectual focus is perhaps best stated by Jim Deetz, when he defines historical archaeology as the "archaeology of the spread of European societies worldwide, beginning in the fifteenth century, and their subsequent development and impact on native peoples worldwide" (Deetz, 1991:1; see also Deetz, 1977).

While Deetz's has become the standard definition of historical archaeology, it has by no means been uncritically accepted by all practitioners. For example, a number of archaeologists who have been influenced by political and social philosophers ranging from Marx and Gramsci to Althusser and Giddens have sought more active definitions of historical archaeology. Sometimes called "critical archaeologists," scholars including Hood (1996), Leone (1995; see also Delle et al., in press), Little (1994), Mangan (1994), McGuire (1992, 1993), Mullins (1996), Orser (1988a, 1990), Paynter (1988, 1989), Potter (Leone and Potter, 1988), and Shackel (1993, 1996) do not necessarily take issue with the underlying assumptions of Deetz's definition, that Europeans spread worldwide and impacted native peoples. Critical archaeologists do, however, seek a definition of the field that encompasses the idea that archaeology is an active component of capitalism, albeit a marginal one (Leone and Potter, 1988:19). Critical archaeologists tend to agree that this sensibility can be extended to both our theories and practices in such a way as to make historical archaeology relevant within the context of late capitalism. To do this, historical archaeologists can first address issues within that political economy; i.e., historical archaeology can be reflexive about the society that produced it, and must understand that analyses of archaeological material say as much about present realities as they do about past societies. Furthermore, and more directly to the point of this study, historical archaeology can contribute to understandings of social processes within capitalism that affect us today as much as they did the subjects of our analyses in the past.

When one examines such historically occurring phenomena as land seizures (Johnson, 1991), African slavery and the concomitant manifestations of racism (Epperson, 1990; Orser, 1988a; Singleton, 1985a, 1991), and gender- and class-based systems of oppression (Gibb and King, 1991; Little, 1994; McGuire, 1991; Mrozowski et al., 1996; Purser, 1991; Scott, 1994a,b; Wurst, 1991), it becomes clear that capitalism is a system fraught with the tensions of social inequality. One goal of a critical historical archaeology can be to understand the processes behind both the creation and maintenance of social inequality in the recent past (Paynter and McGuire, 1991). The creation of social inequality can be analyzed from many angles, including through examinations of the strategies of wealth and power accumulation in play at a given stage of capitalism. The maintenance of such systems has come under increased analysis as the creation of "discipline" among work forces (Delle et al., in press; Hall, 1992; Leone, 1995; Shackel, 1993, 1996), as capitalist elites constantly seek to justify, or legitimate, their social position by rationalizing the nature of inequality (Leone, 1984, 1988a). The creation of labor discipline has been defined by archaeologists as being part of the larger process of "ideology" that naturalizes a status quo in order to mask systems of inequality (Leone, 1995; Miller and Tilley, 1984). These processes are expressed both socially and materially. Because such social phenomena leave material footprints, historical archaeologists have examined the material remains as part of the attempt to understand the development of capitalist social processes (Paynter and McGuire, 1991; Shackel, 1993).

Historical archaeology's contribution to the study of the creation and maintenance of capitalist inequality is focused on the analysis of material culture. Archaeologies of capitalism consider material culture as a crucial element in the negotiation of capitalist social relations. It is through human agency that material culture is given meaning, and through human agency that material culture in turn serves to create new meanings. By endowing specific classes of material culture with exchange value, which may not necessarily reflect its use value, capitalist agency transforms that material culture into a unique social construction: the commodity. Material culture is thus the substance from which comes the creation of commodities and the concomitant development of commodity fetishism.

In recent years, an emerging generation of historical archaeologists has begun to extend the scope of material culture analysis from traditional dating, ceramic chronology and status/ethnicity marking exercises (e.g., Adams and Boling, 1989; O'Brien and Majewski, 1989) to more sophisticated considerations of, for example, the processes of commodification and market creation (LeeDecker et al., 1987; Mullins, 1996; Paynter and McGuire, 1991; Spencer-Wood and Heberling, 1987). Examples of the latter can be read in the work of such scholars as Paul Mullins and Susan Hautaniemi. Mullins (1996) demonstrates how African Americans in Annapolis created and exploited capitalist markets, while Hautaniemi (1992) similarly considers the relationship between emerging mass markets and the negotiation of gender roles at the W. E. B. Du Bois homestead site in western Massachusetts. Both of these projects serve as examples of how material culture studies within historical archaeology can be liberated from traditional boundaries. The important question that such studies ask is not whether status differentiation took place (we know that it did), but what role material culture played in creating and maintaining systems of inequality.

Realizing that social relations within capitalism are inherently unequal, historical archaeologists of capitalism examine the role material culture plays in the construction and negotiation of power and inequality. With this realization has come the effort to understand how social identities and relationships between people of differing social position were materially expressed and negotiated. Following the work of scholars from other disciplines within the social sciences, historical archaeologists have focused this effort on the analysis of race (e.g., Ferguson, 1992; Fitts, 1996; Garman, 1994; Muller, 1994; Paynter, 1990; Paynter et al., 1994), class (e.g., Hardesty, 1994; McGuire, 1991; Paynter, 1989; Spencer-Wood, 1987; Wurst, 1991), and gender identities and relations (e.g., Little, 1994; Scott, 1994b; Seifert, 1991; Wall, 1991, 1994; Yentsch, 1991). Studies in the negotiation of ethnicity have a somewhat longer history within the field (e.g., Ferguson, 1980; Schuyler, 1980).

Some of the earliest historical archaeologies of capitalism explored the wider processes of the economic system. One notable example of such a study is the examination of commodity flow in and out of Silcott, Washington, conducted by W. H. Adams and T. B. Riordan in the late 1970s and early 1980s (Adams, 1976; Riordan and Adams, 1985). A key contribution of this study lay in its recognition that nearly all sites examined by historical archaeologists, even those in the most seemingly remote corners of North America, are connected to wider social and economic networks. These networks, which included the production, distribution, and exchange of commodities, tie localized populations into the wider capitalist system (Riordan and Adams, 1985). Robert Paynter's study of rural western Massachusetts similarly suggests how the historical movement of commodities through space contributed to the construction of social relations on local, regional, and international scales (Paynter, 1982, 1985).

Analyzing the transborder connectedness of disparate populations through the processes of capitalist production and exchange has become the goal of several recent field projects. By comparing colonial sites in differing spatial and temporal contexts, two studies conducted by Orser (1996) and myself (1997, in press) have attempted to demonstrate how historical archaeologists can compare the transnational processes of capitalism. In comparing a nineteenth-century Irish village with a seventeenth-century Brazilian maroon kingdom, Orser suggests the ways in which the lives of people in each of these contexts were tied to the colonial capitalist world order. In my own work, I have attempted to suggest how spatial artifacts can be analyzed to compare the intentions and results of British colonial programs in sixteenth-century Ireland and nineteenth-century Jamaica. Both projects demonstrate that the struggle to control spatial definitions was a key element within global capitalist expansion. As field programs in historical archaeology, these studies further suggest ways in which material elements of colonial settlements can be used to cross-culturally compare the consequences of European expansion.

Such projects reveal the increased attention historical archaeologists have been placing on the international relevance of historical archaeology in understanding the material nature of European capitalist expansion. Interest in examining international questions is growing. Prudence Rice, for example, has recently compared the material culture of wine production in Peru with European antecedents (Rice, 1996a,b; Rice and Smith, 1989). Carmel Schrire (1991, 1992) has recently suggested how soldiers in a South African outpost were connected to the wider colonial world. Ken Kelly (1995, 1997) has examined how contact with Europeans materially affected the negotiation of power in what is now Benin, on the coast of West Africa. Ross Jamieson (1996) has examined how class identity and status were negotiated in colonial Ecuador. As more and more historical archaeologists become interested in international projects, we will gain an increased understanding of how the processes of international capitalism developed and impacted the lives of both the colonizers and the colonized.

SPATIAL ANALYSIS IN HISTORICAL ARCHAEOLOGY

As the processes of capitalism and colonial expansion have resulted in the global expansion of inequality, historical archaeology, with its focus on material culture, can clearly contribute to our understanding of the material basis of that inequality. While many forms of material culture were certainly involved in the negotiation of class, ethnic, race, and gender hierarchies, few have played as ubiquitous a role as space. Within historical archaeology, space has been considered a class of material culture, encompassing what is commonly referred to as "landscape" and "the built environment." Archaeologists have approached the study of space utilizing a variety of analytical scales ranging from the region to the village to the specific site. In order to frame this work in the wider body of literature on spatial theory and methodology, the rest of this chapter reviews the literature on spatial analysis in historical archaeology. In so doing, I will discuss (1) the scales of analysis that have traditionally been used by historical archaeologists (region — village/town/urban — site — house), (2) the development of the historical archaeology of landscapes, and (3) spatial analysis of plantations.

Scales of Analysis

In the early 1980s Paynter (1980, 1981, 1982, 1983, 1985) demonstrated how space on a regional scale could be analyzed to interpret change within the capitalist social order. In his examination of spatial inequality in western Massachusetts, Paynter argued that space should be considered not only a vector of social inequality, but also an active creator of capitalist hierarchies. By controlling the access to the flow of surplus, and by concentrating surplus at entrepôts, regional elites create a system of spatial inequality that supports the maintenance of social inequality (Paynter, 1980, 1982). In the case of western Massachusetts, these processes led to the relative prominence of urban centers located in the Connecticut River Valley, and the eventual abandonment of the so-called hill towns (Paynter, 1985). These phenomena must be understood as processes of development and underdevelopment within a region; the concomitant control of space and commodity production and flow create prosperity for some while others are driven first to rural poverty and eventually to land abandonment (Paynter, 1985).

In her analysis of how space was transformed in late eighteenth-century Catalonia during a theorized shift from feudal to capitalist social relations, Patricia Mangan (1994) studied changes in regional agricultural practices, town plans, and the histories of individual houses to trace shifting relations of production and social reproduction. Like Paynter, Mangan focused her work on a specific region, then argued how changes in the wider political economy affected people within that region, specifically considering the role played by space in redefining social relations.

Among the most influential regional analyses within historical archaeology is Ken Lewis's study of the settlement pattern of the colonial South Carolina frontier (Lewis, 1984, 1985). In this study, Lewis suggested that the spatial processes of European colonialism in southeastern North America are best studied on the regional scale. In investigating what he defined as a frontier region within South Carolina, Lewis constructed a spatial model with which to analyze the processes of colonialism. This model emphasized the importance of the transborder relationship between the colony and homeland, the relevance of transportation routes and technologies to the pattern of colonial expansion, the establishment of a colonial entrepôt from which colonial expansion will spread, the necessity of a hierarchically arranged system of frontier towns and settlements, and the necessity of using a diachronic, evolutionary framework in examining colonial space. This last point is of particular importance, as it recognizes that colonialism is a dynamic phenomenon. As the economic system changes, so too will the spatial expressions of colonialism change. The function, form, and meaning of towns, plantations, and other types of colonial space will change throughout a frontier region through time (Lewis, 1984:1–7, 17–27, 107–113).

One of the key contributions of Lewis's study was the demonstration that historical archaeologists can successfully analyze the diachronic processes of spatial change inherent in capitalist expansion. Equally important is the recognition that synchronic differences in spatial form and function may develop within a specific region. To this end, Lewis defined what can be considered a spectrum of scales of analysis, ranging from the colonial region to the individual entrepôt, town, and nucleated village (Lewis, 1984:111–112).

In examining the spatial dynamics of capitalism at a more refined scale, a number of historical archaeologists have focused their analysis on specific villages, towns, or cities. For example, Ed Hood's consideration of landscapes in Deerfield, Massachusetts (Hood, 1995) suggested how a regional elite intentionally manipulated the town landscape in the attempt to create a false impression of past landscapes. It was the hope of twentieth-century developers in this Massachusetts town to create a sanitized vision of the village's past. This vision served primarily to reinforce the ideology of early twentieth-century capitalism by creating a mythical landscape that reflected modern inequality. By suggesting such landscapes were ancient, twentieth-century capitalists serve to justify their own landscapes of inequality. In contrast, the work of Hood and Reinke in nearby Hadley, Massachusetts (1989, 1990) revealed that the familiar eighteenth-century New England town common was actually a construction of the nineteenth century. This construction served to reinforce the nineteenth-century ideology that pleaded social equality while simultaneously creating inequality. At the core of this deception was the visual demonstration of an imagined past in which people had access to common lands, an access that had been, in truth, strictly limited.

Major contributions at the town scale of analysis have emerged from the excavations undertaken at Annapolis, Maryland, by scholars such as Elizabeth Kryder-Reid, Mark Leone, Barbara Little, Paul

Mullins, Parker Potter, and Paul Shackel. One of the early, and still compelling, arguments to emerge from the Annapolis project stated that in the late eighteenth century, the so-called Georgian elites manipulated their landscapes to reflect their control over both the natural and social orders; by the late 18th century, Annapolitan elites believed that their society — complete with its systems of inequality — was a naturally occurring phenomenon (Leone, 1988a). People like William Paca used created garden layouts as representations of how well they understood the laws of perspective. As Leone put it, "the gardens were regarded as the place to observe and duplicate the precision of the clockwork universe" (Leone, 1988a:255). To control the space of the garden was thus to control the universe.

This line of argument was thoughtfully taken up by Elizabeth Kryder-Reid (1994) in her study of the Annapolis garden site known as St. Mary's, in which she demonstrated that the ability to control the space of the eighteenth-century garden served to reflect the gardener's control of knowledge as well as space, thus they served as powerful media "of the colonial elite to communicate and negotiate their social identity" (Kryder-Reid, 1994:132). In this thought-provoking essay, Kryder-Reid acknowledged the growing acceptance among historical archaeologists that artifacts (including space) both express and produce human behavior. Like the Paca garden, the St. Mary's garden was a public demonstration of power and control by a member of the increasingly powerful capitalist elite, in this case Charles Carroll, a signer of the Declaration of Independence. Men like Carroll sought to publicly demonstrate their control of the laws of perspective; by doing so they created both a spatial and a social order which they controlled.

Another significant contribution to urban-level analysis emerging from the Annapolis project is Paul Shackel's inquiry into the town plans of St. Mary's City and Annapolis, both capital cities of colonial Maryland. This study demonstrated that the baroque town plan of St. Mary's City, which was conceived and implemented in the late seventeenth century, was designed specifically to bolster the class status of Maryland's administrative elite. According to Shackel, the Baroque town plan "may have been conceived by the aristocracy as an expression of their power in the local society in the face of their increasing concern over a growing lower group that had little economic mobility" (Shackel, 1994:88). Simultaneous to the structured town plan was the implementation of racist, segregationist legislation in Maryland; the two phenomena, argued Shackel, served to elevate the aristocratic elite over a newly divided working class, one divided between black and white, between slave and free. A similar set of phenomena can be observed surrounding the transfer of the capital from St. Mary's City to Annapolis in 1694. The Baroque plan of the latter city was conceived so as to place the Anglican Church and the State House on the two highest points in the city. The founders of Annapolis thus created a landscape of power in which these two structures could be seen from any location in the city, constantly reinforcing the authority of both the established Anglican Church and the State (Shackel, 1994:88). The control of space evident in the town plan and architecture of these two capital cities reinforced social and economic hierarchies constructed on social divisions based on class, race, and religion.

Randall McGuire and his students, particularly LouAnn Wurst, have used the concept of town and community landscapes to analyze the changing ideology of capitalism in late nineteenth- and early twentieth-century Broome County, New York (McGuire, 1988, 1991; Wurst, 1991). In examining the differences in the industrial landscapes constructed by Jonas Kilmer in the late nineteenth century and George F. Johnson in the early twentieth century, McGuire demonstrated how the obfuscation of class relations was realized by the manipulation of spatial dynamics. Kilmer, a late nineteenth-century robber baron, constructed his factories and mansion in such a way as to exude power, physically demonstrating the social gulf between Kilmer the capitalist and the workers who labored for him. In contrast, and in reaction to two generations of labor unrest, Johnson constructed an industrial landscape that downplayed social distance; in building his house in the same style as that of his workers, Johnson sought to obscure class differences by spatially manifesting a sliding scale of relative equality. By constructing an industrial welfare system in which company profits were used to subsidize the construction of parks, hospitals, and other public monuments, Johnson attempted to construct an ideology of mutual concern between employers and employees. The purpose behind creating this landscape was to obfuscate class distinction and create an illusion of equality between capital and labor (McGuire, 1991). Such material expressions of the changing ideology of capitalism can be read in the mortuary architecture found in the nearby cemeteries of Broome County (McGuire, 1988; Wurst, 1991). Similar questions concerning the relationship between urban industrial landscapes and the negotiation of power relations have been addressed by Stephen Mrozowski and Mary Beaudry in their long-term study of the Boot Mills industrial complex in Lowell, Massachusetts (e.g., Beaudry, 1989; Beaudry and Mrozowski, 1987a,b; Beaudry et al., 1991; Mrozowski, 1991; Mrozowski et al., 1996).

The question as to how to examine the spatial dynamics of large urban centers has been thoughtfully taken up by a number of historical archaeologists, most notably Nan Rothschild (1990; see also Staski, 1987; Upton, 1992). In her consideration of eighteenth-century New York, Rothschild argued that an effective way to analyze large urban areas is by heuristically dividing a city into subunits, each of which has some internal social cohesion — neighborhoods. Many of the social and spatial processes embedded in modern urbanization can be interpreted at this neighborhood scale of analysis. In analyzing documentary, cartographic, and archaeological material, Rothschild utilized an interdisciplinary methodology to address the intersection between social and spatial dynamics. In particular, she examined the relationship between space and the material construction of and interaction between class and ethnic identities (Rothschild, 1990).

At a yet finer scale of analysis, historical archaeologists have used specific sites, buildings, or clusters of buildings as the focus of spatial analysis. Numerous archaeological and interdisciplinary projects, including studies from the discipline of vernacular architecture, have shown how specific historic buildings can be studied within the realm of spatial analysis. Perhaps the most influential work from this tradition is Henry Glassie's structural analysis of folk housing in Middle Virginia (1975). Glassie used linguistic/anthropological structuralist theory to interpret the act of building construction, while attempting to re-create the thought processes of the constructors. Although Glassie's presentation concentrated on the mental processes of house construction and not on how these buildings helped to create inequality in the landscape, his work, as well as that of his contemporaries (e.g., Carson et al., 1981; Deetz, 1977; Neiman, 1978) firmly demonstrated that specific buildings should indeed be considered by historical archaeologists concerned with spatial analysis. More recent generations of historical archaeologists have incorporated such ideas into studies beyond the Chesapeake and Virginia (e.g., Armstrong and Kelly, 1990; Baram, 1989; Johnson, 1993; Orser, 1988b).

Perhaps the most eloquent contribution from the interdiscipline of vernacular architecture to spatial analysis in historical archaeology comes from Dell Upton. In his consideration of Chesapeake tobacco plantations, Upton argued that the most evident material expression of power on tobacco plantations are the spatial relationships between the landscapes of the planters and the landscapes of the slaves (Upton, 1986, 1988). Chesapeake tobacco plantations were generally isolated and self-contained communities, which Upton metaphorically likened to a village, with the planter's house serving as a "town hall" or center of power. As a center of administrative and economic power, the planter's house would be elevated above other buildings, on a hill for instance, and would be physically separated from the surrounding countryside by systems of terraces and fences. Plantation houses were intentionally constructed to dominate the landscape. One firsthand account reported that the plantation house could be seen from a distance of 6 miles. More to the point, by the late eighteenth century, plantation houses, like Jefferson's Monticello, were constructed to command a view of the landscape as well. The effect of the creation of such a spatial division was that both slaves and masters shared an attitude about the possession of space. While this afforded some measure of spatial proprietary rights to the slave, such a shared consciousness, if it did indeed exist, reified the work discipline of the plantation. The space that both the planters and slaves acknowledged as the domain of the slaves included the work spaces: bakeries, blacksmith shops, fields, and so on (Upton, 1988).

Landscape Archaeology

Many historical archaeological studies that address spatial questions have recently been categorized under the rubric of "landscape archaeology." When one considers the range of work that has been catalogued under this heading, it becomes clear that no one specific definition of "landscape archaeology" has been adopted by historical archaeologists. There are, however, several elements that most landscape studies share. First and foremost, landscape archaeologists examine relatively large spatial units — historical landscapes. Most landscape archaeologists would agree that the spaces that people build and occupy are endowed with multiple meanings, meanings that change with social situations, and that change through time. Finally, most landscape archaeologists would contend that material landscapes both shape and reflect social relations.

This particular field of inquiry became influential within historical archaeology in the mid-1980s. Three important events mark the beginning of the florescence of landscape studies in historical archaeology. The first of these was the incorporation of garden archaeology into larger historical archaeology programs, such as that led by Mark Leone in Annapolis (Leone, 1984, 1988a,b; Leone and Shackel, 1990). While formal gardens had been of interest to archaeologists at least since the early 1970s (e.g., Noël Hume, 1974; Kelso, 1984), the mid-1980s saw an explosion in interest with garden archaeology. As Beaudry (1996) has recently argued, interest in garden archaeology was stimulated by both intellectual and logistical factors, including an increasing interest in reconstructing historic gardens as part of larger historical preservation projects, and the resulting ability to secure funding for the examination and reconstruction of historical gardens. Landscape archaeology in gardens is now fairly common, with studies having been completed in, for example, urban gardens in New England (Mrozowski, 1987), slave gardens in the Caribbean (Pulsipher, 1994), plantation gardens in Virginia, Maryland, and the Carolinas (Kryder-Reid, 1994; Sanford, 1990; Trinkley et al., 1992), Chinese gardens in the mountains of Idaho (Fee, 1993), formal gardens in New Jersey (Metheny et al., 1996), and the gardens of presidents in Tennessee and Virginia (McKee, 1996; Pogue, 1996).

The second and third key events that stimulated interest in landscape archaeology transpired at academic conferences held in 1986 and 1987. The former was a 1986 conference on landscape archaeology sponsored by the University of Virginia and the Thomas Jefferson Memorial Foundation. This conference brought together scholars examining historical landscapes, primarily in the Tidewater, but also from northeastern North America and classical Italy. Many of the conference papers were published in a volume edited by William Kelso and Rachel Most (1990). The latter event was a symposium organized by Faith Harrington for the 1987 meeting of the Society for Historical Archaeology: "The Archaeological Use of Landscape Treatment in Social, Economic, and Ideological Analyses." This symposium brought together scholars concerned with spatial questions on a variety of different landscape forms, including urban houselots, industrial spaces, rural vernacular landscapes in New England, Tidewater formal gardens, and California ranches. Three of the six papers were published in the journal *Historical Archaeology* in 1989 (Beaudry, 1989; Harrington, 1989; Praetzellis and Praetzellis, 1989).

The three published articles were accompanied by brief remarks written by three influential scholars (Kelso, 1989; Leone, 1989; Rubertone, 1989), each of whom commented on the increased importance of landscape archaeology to the broader field of historical archaeology. For example, Leone (1989:45) stated that the symposium papers "represent a significant initiative in historical archaeology." The opening line of Rubertone's comments suggested that landscapes "are among the most intriguing artifacts studied in historical archaeology" (Rubertone, 1989:50). Kelso (1989:48) was even more laudatory in his statements concerning the potential for landscape archaeology, arguing that "[i]f ever a symposium had a chance to demonstrate that despite a staggering diversity of data, archaeological research is capable of reaching a higher, common, and useful plane of cultural interpretation, then it was indeed the 1987 . . . symposium on landscapes." Although these three scholars differed somewhat in their opinion as to whether the negotiation of power relations was indeed embedded in historical landscapes, they agreed that landscape archaeology had become an extremely important focus of historical archaeology. The success of the symposium and the appearance of the papers and comments in historical archaeology's primary publication venue underscore the status attained by landscape studies within historical archaeology by the end of the 1980s.

The momentum gained by landscape studies during the late 1980s has been sustained through the mid-1990s. Numerous articles addressing landscape issues have appeared, and several important anthologies have brought together the results of some of the better landscape projects. Of particular note are three collections of scholarly articles: *Earth Patterns: Essays in Landscape Archaeology*, based on the landscape archaeology conference discussed earlier, and edited by William Kelso and Rachel Most (1990); *The Archaeology of Garden and Field*, edited by Naomi Miller and Kathryn Gleason (1994); and *Landscape Archaeology: Reading and Interpreting the American Historical Landscape*, edited by Rebecca Yamin and Karen Bescherer Metheny (1996). An interesting popular volume on the historical archaeology of space and landscape has also recently appeared, titled *Invisible America: Unearthing Our Hidden History*, edited by Mark Leone and Neil Asher Silberman (1995).

The Spatial Analysis of Plantations

As this book is not only a study in spatial analysis, but also a work of plantation archaeology, I think it appropriate to frame my contribution to the literature on the historical archaeology of plantations. Analyzing the material culture of plantations has been a focus of historical archaeologists since the very earliest days of the field. Plantation archaeology is as old as the leading North American professional organization in historical archaeology, having been launched within a few months of the founding of the Society for Historical Archaeology (Singleton, 1985b; South, 1994). The key figure in the creation of plantation archaeology is Charles Fairbanks, one of the "pioneers in historical archaeology" identified by Stanley South (1994). Singleton (1985b:1) has suggested that plantation archaeology can trace its origin to Fairbanks's 1968 test excavations of slave cabins in Duval County, Florida; the SHA was founded just a few months prior to these excavations, in 1967 (South, 1994:vii). Thus, the establishment of plantation archaeology nearly coincided with the official institutionalization of historical archaeology as an academic field.

Plantation archaeologists have examined sites that date to the days both before and after the abolition of slavery. The subjects of inquiry have included plantations that produced cotton, sugar, rice, indigo, and most recently coffee. The geographic range of plantation archaeology encompasses the Gulf Coast of the United States from Louisiana to Florida, the eastern coast from Florida to Maryland, the Georgia, North Carolina, and Virginia piedmont, Tennessee, and a number of Caribbean islands, including Barbados, St. Eustatius, Montserrat, Puerto Rico, and Jamaica.

Not surprisingly, the scholars who have produced the canon of plantation archaeology literature come from varied intellectual traditions ranging from the cultural historicism of Noël Hume (1982), the structuralism of Deetz (1993), the ethnohistoricism of Handler and Lange (1978), to the processualism of Armstrong (1990). The particular foci of each of these studies were products of specific paradigms that emerged from an archaeological discourse current at the time in which the particular knowledge was produced. Similar influences in the late 1970s and early 1980s produced a number of plantation studies that attempted to define status on plantations (e.g., Moore, 1985; Otto, 1977, 1984), a key concern of historical archaeologists at that time. Simultaneously, the concern with defining the relationship between material culture and ethnicity influenced the direction of other studies (e.g., Babson, 1990; Ferguson, 1980). In the late 1980s and into the 1990s, with the emergence of the post-processual critique, some plantation archaeologists shifted their focus to the construction of power and power relations (see in particular Orser, 1988a,c, 1991).

Yet another late-1980s/early-1990s trend is the utilization of the methodologies of landscape archaeology to analyze plantation spaces (e.g., Armstrong and Kelly, 1990, 1991; Delle, 1994; Hudgins, 1990; Kelso, 1990; Lucketti, 1990; McKee, 1996; Orser, 1988b; Pogue, 1996). As compared with the number of studies concerning relative status and ethnic life ways of the laboring population on plantations, relatively few studies have focused specifically on the spatial analysis of plantations, at least beyond the formal gardens constructed by the elite plantation owners (Pulsipher, 1994; for a good review of the archaeology of slave life, see Singleton, 1991). The most comprehensive spatial analysis of plantations to date is Orser's consideration of Millwood, a South Carolina cotton plantation that operated during both the antebellum and postbellum periods (Orser, 1988b,c; Orser and Nekola, 1985). As Orser's spatial analysis of an American plantation during the transition away from the use of slave labor is both theoretically and substantively similar to the project discussed in this book, I think it worthwhile to consider the major points of his argument.

According to Orser's presentations, the planter-owner of Millwood Plantation was one James Edward Calhoun, a well-connected member of the southern elite (his father, John E. Calhoun, was a U.S. Senator; his cousin and brother-in-law, John C. Calhoun, served at various times as U.S. Representative from South Carolina, U.S. Secretary of War, Vice President of the United States, U.S. Senator, and Secretary of State). The local oral tradition suggests that during the antebellum operation of the plantation a slave driver inflicted corporal punishment that James Calhoun would supervise (Orser, 1988b:36). As early as 1833, Calhoun began the system of plantation tenancy that would become the dominant system at the plantation after the Civil War.

The defeat of the Confederacy forced the planter elites to reorganize the system of labor exploitation; although the planters were confused and frightened about their future, there was a general resolve not to transfer ownership of the plantations to the emancipated slaves (Orser, 1988b:49). As a result, several strategies of labor extraction were established, including farm wage labor and systems of time and crop sharing (Orser, 1988b:45–51). During the years following the war, the South Carolina elite created new systems of discipline both at the state level through the establishment of a so-called "Black Code" and at the level of the individual plantations. The Black Code, established in 1865, legally restricted the movements and labor of black workers and their families. The individual planters entered into contracts with laborers, which again sought to construct a labor discipline that would be as beneficial to the planter as possible. Orser reports that on one Louisiana plantation, any laborer not reporting to the fields by the morning bell would be docked one quarter of a day's pay. Pay could also be docked for arriving late after lunch or for acts of insubordination (Orser, 1988b:51, 1991:41).

Both freedmen and planters disliked the wage labor system, the freedmen because of the rigid work regimen imposed on them and planters because of the lack of direct control they had over the laborers (Orser, 1988b:53). The dialectical conflict between the desire on the part of freedmen to own land and thus to control their own destinies and the desire of the planters to retain their social and economic power and thus their unwillingness to sell land resulted in the development of the plantation tenancy system, which allowed black farmers access to land that would remain the property of the planters-turned-landlords (Orser, 1988b:55). Two classes of tenancy developed: renting and sharecropping. There was a legal distinction made between renters and sharecroppers vis-à-vis the crop they produced. Renters were the legal owners of their crops while sharecroppers were not (Orser, 1991:43).

The restructuring of labor following emancipation resulted in the restructuring of the plantation settlement form. Unlike antebellum plantation settlements, which put a premium on direct surveillance of the laborers, postbellum plantation settlements tended to be dispersed; each farmer, whether renter or sharecropper, lived with his or her family near the fields that they worked. Significantly, where sharecroppers made up the bulk of the plantation tenants, the settlement form more closely resembled antebellum forms in that barns, sheds, and other outbuildings would be located near the planter's settlement complex. This provided the planter-landlord with the capability to directly supervise the sharecroppers. When a farmer wanted to use, for example, a mule or a plow, that farmer would be forced to approach the planter's house, under the direct surveillance of the landlord. Likewise, by locating the barns within sight of the house, the landlord could maintain direct surveillance over the harvested crop. Renters were the objects of relatively less surveillance because the landlords had less direct interest in the renters' crop (Orser, 1988b:92).

Orser's approach to the analysis of plantation space has effectively utilized a Marxist theory and methodology to understand the politics of space on a plantation during a critical transition within American capitalism. His work has demonstrated not only how historical archaeology can interpret space as material culture but, more specifically, how a planter elite conceptualized a new set of social relations following emancipation. He also has suggested how elites used space and spatial relations to implement and reinforce the new social order enabling them to remain in control of land and production. Finally, Orser has illustrated how the dialectics of spatial understanding were negotiated between the laborers and the landlords, resulting in resistance to a new social order that promised freedom but delivered a new system of oppression. In effect, Orser's analysis considers the negotiation of material culture, including the spaces in which people existed, to be embedded in the larger negotiation of power and class relations, overarching social phenomena that must be considered in plantation studies (Howson, 1990; Orser, 1988a,b). As such, Orser constructed a model with which diachronic spatial and social changes on plantations can be analyzed.

CONCLUSION

This book considers a case study theoretically similar to that analyzed by Orser. Just as the shape of American plantations changed in the years surrounding the end of slavery, so did the shape of Jamaican plantations change a generation earlier. The changes that occurred in Jamaica were the result of global changes in the organization of capitalism. This book considers the nature of spatial changes that occurred in Jamaica through an analysis of coffee plantations in the Blue Mountain region. In doing so, I attempt to discern how global changes were manifested locally, that is, how changes in the capitalist mode of production reverberated in the coffee-growing district of eastern Jamaica.

Although there is a rich historical record of both the British West Indies at large and Jamaica in particular, to date most of the studies undertaken on the history and historical archaeology of this region have concentrated on the development and/or fall of the sugar industry, and the plantation political economy that developed to support it. Literally dozens of monographs and articles have been written about sugar and sugar plantations (e.g., Armstrong, 1990; Craton, 1978; Craton and Walvin, 1970; Deerr, 1950; Delle, 1989, 1994; Dunn, 1972; Handler and Lange, 1978; Mintz, 1985a; Pulsipher, 1994; Sheridan, 1974; Tomich, 1990). In stark contrast, very little has been written about coffee plantations in Jamaica, or throughout the British Caribbean for that matter. No modern published monographs dedicated to the history of Jamaican coffee plantations exist, and there are but few other publications dedicated to this particular subject (e.g., Higman, 1986; Montieth, 1991). One of the goals of this project is to continue to explore the path broken by these scholars by closely examining the development of the infrastructure of the early Jamaican coffee industry.

Despite the completion of a C.R.M. report on coffee plantation sites in Puerto Rico (Joseph et al., 1987) and a few occasional papers presented at meetings of the Society for Historical Archaeology (e.g., Delle, 1995; Joseph, 1995; Reeves, 1996), little has appeared in the published literature of historical archaeology on Caribbean coffee plantations. In framing this study, therefore, I find myself with few studies with which to directly compare my work on this topic. The state of historiography for coffee plantations in the French Antilles is a little better than that for the British Caribbean. Michel-Rolph Trouillot (1982) and David Geggus (1993) have each published articles on coffee production in St. Domingue; Trouillot has published in both English and Spanish. Recently, Trouillot (1993) published a short article discussing coffee's role as a secondary crop in the Americas. Unfortunately, little else has yet been published on eighteenth- and early nineteenth-century Caribbean coffee plantations; most of the literature on coffee plantations in Latin America concentrates on the industry in the later part of the nineteenth century.

This book examines the processes inherent in the negotiation of landscapes and space during a period of significant regional change within global capitalism, specifically on coffee plantations in Jamaica's Blue Mountains. As the following chapters unfold, we will see how the social spaces of nineteenth-century Jamaican coffee plantations were manipulated, negotiated, and changed in response to changes in the organization of global capitalism. As was the case in mid-nineteenth-century South Carolina (Orser, 1988b), early nineteenth-century Jamaica was experiencing a period of dramatic socioeconomic restructuring. A series of global crises within capitalism precipitated a coffee boom that lasted from, roughly, the late 1790s through the 1820s. After the middle of the 1820s the Jamaican coffee industry began to decline, facing yet another transitional crisis with the abolition of slave labor. This study explores how Jamaican elites manipulated these transitions in global capitalism in the attempt to reinforce their elevated social position. I consider the material and documentary remains of the coffee-producing region known as the Upper Yallahs or Blue Mountain region of Jamaica, in order to discuss how formerly "undeveloped" space was transformed into productive (i.e., commodity producing) space; how the planter class intentionally manipulated space, both prior to and following emancipation, as part of the strategy to create, maintain, and legitimate social and economic inequality; and how the working people of Jamaica resisted, both reactively and proactively, these changes.

To this end, I analyze space on the scale of both the region and the individual plantation. This book focuses on a specific region, the Yallahs River drainage of the Blue Mountains of Jamaica, considering how changes in the political economy of capitalism effected change within that region. It explores how the meanings of specific landscapes on coffee plantations — the dominant spatial form within the study area — were negotiated between elite planters who owned and managed the plantations, and the workers who actually produced the commodities.

2 Theoretical Background
Capitalism, Crisis, and Social Space

There is no man who has seen more of slavery than I have, and there could be no worse institution; but there is this about it, that in a state of slavery, the planter may go on from day to day and year to year upon credit, working the estate by the labour of the slaves; but in the present state of society he cannot move an inch unless his pocket is full of money. In a state of slavery there was no money outlay; in a state of freedom there is a constant drain upon the planter.
—Alexander Geddes, Jamaican coffee planter, 1848

INTRODUCTION

During the first half of the nineteenth century, coffee was a major component of Jamaica's export economy, ranking second only to sugar in importance within the island's plantation system. In comparison with sugar, large-scale coffee production was introduced to Jamaica relatively late; it was not until the opening decade of the nineteenth century that coffee production was attempted in earnest. Throughout much of the seventeenth, and most of the eighteenth century, the Jamaican economy was dominated by monocrop sugar production. Although some cotton, indigo, and pimento production had been attempted, these crops never amounted to much in comparison with sugar. Coffee was the first secondary crop of importance to be grown in Jamaica, and (prior to the modern Blue Mountain coffee boom) really only succeeded for a few short decades. Sugar production had so dominated the Jamaican economy that coffee production became relevant only in reaction to a series of crises within the West Indian sugar-based political economy. As the coffee industry was both extensive and short-lived, a spatial analysis of coffee plantations is particularly informative when addressing the relationships between the intentional design of space, the actual construction of space, and the negotiation of social relationships that occurred within these spaces.

In subsequent chapters I will consider three phases of the social history of space on Jamaican coffee plantations: how coffee plantations were imagined before they were actually constructed; how coffee plantations were initially constructed and negotiated in Jamaica under the slave system; and how coffee plantations changed following the abolition of slavery. The construction of this tripartite historical schema is based on an interpretation of two shifts within the regional political economy of Jamaica, both of which can be defined as resulting from periods of crisis within the wider colonial world economy. In both cases new definitions of space were used in the negotiation of new relations of production, and thus form the core of this book's analysis. Because this analysis requires my using the concepts of "capitalism," "crisis," and "social space," this chapter is dedicated to the definition of these concepts in terms of a general theory on the structure of capitalism, and a "middle-range" theory that links observable spatial phenomena to this general theory.

GENERAL THEORY: CAPITALISM AND ITS CRISES

This study is based on a political economic theory of colonialism, distilled from world systems theory, which contends that, historically, members of certain elite classes based in Europe extracted wealth at the expense of members of other classes, both in Europe and in European-controlled colonies. The social process through which this wealth extraction is negotiated is capitalism, a class-based political economy. A remarkable feature of capitalism is its historical flexibility. Many political economies have collapsed in the face of severe socioeconomic crisis — one need only think of the recent collapse of state socialism in Eastern Europe and the former Soviet Union to understand this point. The structures of capitalism, on the other hand, have had the historic ability to be transformed rather than collapse in the face of such crisis. These various historical transformations have often resulted in the elevation of one or more sectors of the political economy that successfully challenged the hegemony of previously dominant sectors, i.e., producers of certain types of commodities or newly organized classes may become ascendant within the system. However, despite such periods of social restructuring, the overarching logic of the system has remained intact.

Defining Capitalism

Capitalism is a political economy characterized by a type of stratified social structure in which human relationships are defined by membership in and allegiances to social classes. Membership in capitalist social classes is defined by an individual's ability and opportunity to accumulate wealth. These abilities are directly related to the control of strategic resources through individual ownership of private property, including the tools, knowledge, and raw materials required to produce commodities for exchange. Under the capitalist system a small elite owns and maintains control over these means of production, while a majority of the population is required to work for these owners — either by having their labor power coerced from them through the institution of slavery, or through the necessity of selling their labor power to the owners for a wage. The social realities revealed in this latter statement have resulted in a variety of theories of history and value-laden definitions of capitalism which in turn have resulted in a variety of historical interpretations of the events of the nineteenth century. As I argued in the previous chapter, historical archaeology may be best defined as an archaeology of capitalism; as such, those pursuing historical archaeology should think closely about the definition of capitalism they choose, and thus the definition of the archaeology of capitalism that they practice. In this book, I use a definition of capitalism derived from the work of Tom Patterson, Eric Wolf, Immanuel Wallerstein, Robert Paynter, and Charles Orser.

The final chapter of Patterson's recent textbook on the historical development of civilizations is dedicated to a discussion of the archaeology of capitalism, in which he presents a basic definition of capitalism. According to this definition, capitalism is an economic system concerned with the production and sale of commodities in markets (Patterson, 1993:350). Patterson distinguished commodity exchange under capitalism from other types of exchange, noting that capitalist exchange differs qualitatively from "precapitalist" exchange. The key distinction between precapitalist and capitalist exchange is in the cultural construction and understanding of value. In precapitalist economic systems it is the use value of a good that is of consequence, while under a capitalist system the exchange value, that is, the price a commodity can fetch on a market, is prioritized over its use value. Under capitalism, "value is continually being transformed from money into commodities and then back into a larger sum of money; value becomes capital by virtue of the fact that it is involved in the process of expanding, of creating extra value" (Patterson, 1993:350). Following Marx, Patterson argued that it is only human labor power that can create surplus value (Marx, 1967:46ff; Patterson, 1993:350). For capitalists to accumulate wealth, they must commodify labor.

This process of labor commodification is crucial for an understanding of capitalism at large, and the specific processes in effect in nineteenth-century Jamaica. Part of the crisis that occurred in the British Atlantic world during this era was the redefinition of labor as a commodity. Under slavery, labor is easily understood as a commodity, as the providers of labor — enslaved Africans — are themselves defined and treated as commodities to be exchanged in markets. However, as the nineteenth century progressed, and the slave trade and eventually slavery itself were abolished by Britain (in 1807 and 1834 respectively), the nature of labor commodification was transformed. In order to purchase commodities themselves, freed laborers were required to engage in wage labor, by which they sold their labor power to agrarian capitalists. Hence, part of the crisis of capitalism in the nineteenth century involved the transformation of the relationship between labor and capital in such a way as to commodify labor, while simultaneously decommodifying the laborer. What comes to be bought and sold is not the laborer him- or herself, but rather the laborer's ability to work — his or her labor power (Marx, 1967:193).

In his landmark study *Europe and the People without History*, Eric Wolf expounded on the changing nature of the relationship between labor and capital. He considered labor as a social process, that is, labor is "always mobilized and deployed by an organized social plurality" (Wolf, 1982:74). Thus, in order to understand capitalism, as well as other strategies of social organization, one must understand how human beings organize and mobilize labor and production.

Wolf utilized the concept of the mode of production to compare how distinct social systems have organized production. According to his analysis, a mode of production is "a specific, historically occurring set of social relations through which labor is deployed to wrest energy from nature by means of tools, skills, organization and knowledge" (Wolf, 1982:75). He further suggested that the concept of "mode of production" is an abstraction that allows scholars to reveal the political economic relationships that underlie, orient, and constrain social interaction (Wolf, 1982:76).

Under the capitalist mode of production, the commodification of labor is dependent on the alienation of labor from the means of production, that is, the tools, equipment, raw materials, and knowledge required to produce commodities for exchange. Once capitalists have gained control over these means of production, laborers will be left little choice but to sell their labor power on the market. Thus, in order to complete the transition from a mode of production based on the direct coercive control of the physical beings of laborers as existed under slavery, to a system by which capitalists could buy the labor power of alienated workers, elites needed to establish and maintain control over the means of production. Furthermore, for capitalists to maximize accumulation, they must continuously alter the conditions of production to maintain an increasing rate of profit. For example, if one set of capitalists were able to decrease the cost of production by decreasing the cost of labor, other capitalists would need to follow suit, or face the probability of being undersold in commodities markets and hence brought to ruin. Such manipulations can alter the specific relationships between capitalists and labor, while preserving the control of the means of production enjoyed by capitalists.

I should note that Wolf did not distinguish different classes of capitalist modes of production. According to Wolf there "is no such thing as mercantile or merchant capitalism. . . . There is only mercantile wealth. Capitalism, to be capitalism, must be capitalism-in-production" (Wolf, 1982:79). Wolf argued that the capitalist mode of production emerged only when *both* the means of production *and* labor power became commodities for sale in markets. Thus, he argued that capitalism did not develop until the latter part of the eighteenth century (Wolf, 1982:298).

Although there is a significant difference between what I call the mercantile capitalist mode of production and the competitive capitalist mode of production, both were historical variants of capitalism. The mercantile mode as historically experienced within British capitalism was characterized by a large sector of the laboring population being commodified as labor; Africans were enslaved, and to an extreme degree, were not permitted to sell their labor power for wages. Under the slave regime in Jamaica (as elsewhere), the planter class claimed ownership not only over the laborers themselves as chattel, but also over their labor power. The laborers had very limited opportunities to sell their labor power in markets; the planters, however, did sell the labor power of the enslaved peoples in markets. For example, at Radnor plantation in the Blue Mountains, one of the estates to be discussed in detail in later chapters, the management would rent out the labor power of the enslaved laborers to other estates, in effect selling their labor power. This phenomenon seems to run counter to the argument that only "when the stock of wealth can be related to human energy by purchasing living energy as 'labor power,' offered for sale by people who have no other means of using their labor power to ensure their livelihood; and only when it can relate that labor power to purchased machines . . . only then does 'wealth' become 'capital'" (Wolf, 1982:298). In this model, capitalism thus can only exist in industrial contexts without the use of slave labor.

Fully reconciling the contradictions between slavery and classic definitions of capitalism, such as Wolf's, is a task that is beyond the scope of this particular study. However, because the transformations in space that occurred in nineteenth-century Jamaica were in part a result of the reconciliation of a slavery/wage-labor dialectic within the British Empire, it is worth considering this point further. The key to understanding the nature of this contradiction is the nature of the slave as laborer, provider of labor power, and a commodity purchased by the capitalist as a form of human capital. During the period of slavery in the New World, both the labor power and the corporeal bodies of enslaved workers were considered commodities for sale in the labor markets of the world economy. How then does one define the mechanism by which labor power is transformed into capital?

Wallerstein (1979:17, 216–217) addressed this question by arguing that from the perspective of the slave masters, enslaved laborers were indeed paid a wage based on the amount paid for purchasing (or cost of raising) the individual, the cost of care during youth and old age, and the cost of "policing him [sic]" (Wallerstein, 1979:217). As such, capitalism existed where the social relations between the master and the slave were mediated by a metaphorical wage defined by these expenditures. However, Wallerstein's metaphorical wages do not seem to apply to Jamaica, where the actual cost of providing food and care for the elderly or young often fell to the enslaved population itself. Although Jamaican planters did occasionally supply special provisions for the laborers at holidays, daily food was grown by the enslaved themselves in provision grounds. Many elderly people, and children from the age of 5, were expected to be part of the laboring population on an estate. As far as policing goes, much of the surveillance that occurred on the plantations was executed by enslaved drivers and colluding informers. When a slave was purchased, money was exchanged between the planter and the slave merchant. Very little of the metaphorical wage actually was paid to the enslaved worker.

A more compelling argument made by Wallerstein considered slavery as a form of labor extraction compatible with wage labor within a larger capitalist world economy. Wallerstein (1979:147–148) argued that capitalism is not based solely on free labor and land, but is a mode of production that combines proletarian labor and commercialized land with other forms of labor and land relations. According to this argument, whichever form of labor extraction is most profitable within a region of the world economy will be the form that dominates (Wallerstein, 1974:86–88, 1979:178). In the periphery of the capitalist world system, slavery was a more cost-effective labor system so long as a ready supply of human beings was available, and the majority of the work required was menial and nonskilled (Wallerstein, 1974:88). Wolf (1982:87) agreed that while slavery does not constitute the basis of a distinct mode of production, "it has played a subsidiary role in providing labor under all modes," particularly where "output is dependent on a maximization of labor, with minimal deployment of tools and skills."

It is clear that while both forms of labor extraction rely on the exchange of the products of labor power, slave labor differs considerably from the use of wage labor. Under a slave system the worker does not even nominally control the exchange of labor power within a labor market. A second party, the capitalist-planter, is given the right to own not only the results of labor power, but also that labor power itself. The slave is not free to sell the products of labor or labor power itself; the slave exists as perhaps the most alienated individual within a capitalist system. Wage laborers, on the other hand, while alienated from the means of production, are nominally given some measure of choice as to where and to whom they will sell their labor power. The simultaneous existence of these two labor extraction strategies created contradictions within the capitalist mode of production; contradictions that contributed to the shift from mercantile to competitive capitalism.

Wolf argued that the capitalist mode of production is necessarily dependent on a hierarchical division of society based on class. Paynter's exposition on this component of capitalism is particularly germane to this study, as he consciously endeavored to theorize how material culture contributes to the negotiation of capitalist social relations (Paynter, 1982, 1985, 1988; Paynter and McGuire, 1991). In his appraisal of why material culture changed so dramatically in the industrialized nineteenth century, Paynter argued that the key to understanding the explosion in the quantity and variety of consumer goods lies in the relationships between classes. According to Paynter's (1988) synthesis of capitalist class modeling, capitalists own the means of production, but require labor to produce objects. As workers are alienated from the means of production, they are forced to sell their labor power to capitalists in return for a wage. The result of this social relationship is the production of commodities that are made by the workers, but are appropriated by capitalists who claim ownership over the commodities. As capitalists compete with one another to sell their commodities, more and more are produced, resulting in the exponential increase in the volume and types of consumer goods that historically appeared in the nineteenth century. According to Paynter's argument, the social distinction between the alienated worker who produces the commodities and then must buy them back from the appropriating owner is the basis for class division in capitalism.

A shortcoming in much of Marxian thought is a general failure to explain the significance of slave labor to the circulation of capital. Marx argued that class relations between capital and labor are mediated by wages; it is through the negotiation of wages that surplus value, the determinant of capital accumulation, is constructed. In Marxian thought, there is a critical relationship between wages and exchange value. Paynter summarized this line of argument: "The exchange of labor power for wage, the production of commodities and profits, constitutes the circulation of capital" (Paynter, 1988:413). How, then, does one place coerced, nonwage labor into the circulation of capital?

In early nineteenth-century Jamaica, commodities and profits were indeed created without the benefit of wage labor. It would seem that the key element in the negotiation of capitalist class relations is the cost of harnessing labor to produce commodities, whether through the negotiation of wages or through the maintenance of a slave labor force. In either case, the process of work transforms capital into material culture. Under a slave regime, that capital is invested into labor as a means of production; human bodies are bought and sold as implements for the construction of commodities. This is a class relation in as much as money is spent for labor; whether that money is spent as wages or for chattel does not change the fact that these are class relations. It does, however, change the nature of those class relations.

In an earlier article, Paynter (1985) outlined a schematic of capitalist class relations germane to this study. According to Paynter's adaptation of world systems theory, capitalist class relations can be conceptualized as sets of behaviors that exist between at least three different interest groups. Primary producers are the people whose labor is exploited; as the term indicates, these people are directly involved in the production of commodities. The class of regional elites are those in direct control of the means of production. In colonial situations like nineteenth-century Jamaica, these people are the expatriots and creoles who actually live in the producing regions. A group among the regional elites has direct supervisory control over the primary producers. Core elites are those who remain in the metropolitan area and control the flow of surplus. In the Jamaican case, this group included not only absentee plantation owners, but also Parliament, the Lords of Trade, and so forth, i.e., those in control of the British political economy residing in Great Britain (Paynter, 1985; Wallerstein, 1980). This model is useful in conceptualizing the class structure in Jamaica, as the class of primary producers need not be wage-based laborers, but could include enslaved laborers. What is important to the analysis of class relations in Jamaica is the idea that the lives of people within this group were manipulated by both regional and core elites in the constant attempt to accumulate wealth.

Orser has attempted with great success to redefine how archaeologists perceive and interpret the nature of class interaction on slave-based plantations. According to Orser (1988a), many scholars rationalize the contradictions inherent in a capitalist system based on slave labor by drawing an analogy between New World slavery and the Hindu caste system. He argued that archaeologists are amiss in drawing this analogy; rather, relationships on capitalist plantations should be conceived as being based on differential levels of social power exercised by the planters and the enslaved. In this model two distinct classes exist on plantations: owners in the persons of planters and their agents, and direct producers in the persons of the enslaved laborers. Orser's argument is an important one. Rather than merely interpreting slavery as a system through which the labor power of one class was stolen by another, one should consider that plantation class relations are mediated through the exercise of social power, reified by a racist ideology. This power is used to coerce labor to work rather than allow people access to markets in which labor power could be sold. By considering this model, Orser introduced the idea that slavery can be understood in capitalist terms as the expression of unequal power relations that exist between classes. Social stratification is thus created and maintained by the members of one class who control access to material objects, property, and space itself.

In summary, capitalism is a social formation based on the construction of social classes. The more powerful classes control the production and distribution of commodities in markets, and manipulate social relations in order to maximize profit by continuously increasing their rate of profit. The working classes are alienated from the means of production, and are required to labor for someone else, either as an enslaved and dehumanized means of production, or as wage laborers alienated from the means of production. The capitalists, who control access to the means of production, use the power inherent within this social structure to accumulate wealth, power, and prestige.

The Idea of Crisis

A central proposition of this study contends that changes within capitalist social relations will result in changes in the spatial organization of production. In order to understand how these changes transpired, it is necessary to consider why such changes occurred in the first place. Capitalism is an extremely flexible social formation, which is continuously changing. During historically occurring periods of social stress, brought on by, for example, the exploitation of new resources or the development of new technologies, the operating logic of capitalism changes. Such moments are often referred to as crises. During such moments of stress, the class basis of social relations remains, but the composition of social classes, and the specific logic of capital accumulation may change. In short, a crisis in capitalism is a period when the circulation of capital is disrupted. Such disruptions most often occur when significant sectors of the political economy are unable to maintain an increasing rate of profit (Wolf, 1982:299). Crisis can be experienced by individual capitalists or their companies, by entire sectors of the economy (e.g., the sugar industry or the coffee industry), regions of surplus extraction within the global system, or the entire system itself. Crises not only often affect those who control the circulation of capital but, more significantly, disproportionately affect those whose labor power is involved in the production and distribution of the commodities experiencing crisis. When the disruptions in the circulation of capital are pervasive and persistent, industries, sectors, or the capitalist system as a whole adapt through a restructuring of those elements within the forces and/or relations of production that can be blamed for the disruption. Thus, in recent times, organized labor, federal welfare programs, and protective trade tariffs have come under attack by those hoping to redirect the circulation of capital.

For the purposes of this study, I will define as crises only those periods of disruption that result in significant social restructuring. Following O'Connor (1987:3), I define a crisis period within capitalism as a historical moment during which the subjective action of individuals serves to redefine the nature of social relations within the capitalist order; it is at such moments that decisions made by capitalist elites in effect re-create the capitalist mode of production. As O'Connor (1987:3) has phrased it, crisis "is not and cannot merely be an 'objective' historical process. . . . 'Crisis' is also a 'subjective' historical process — a time when it is not possible to take for granted 'normal' economic, social and other relationships; a time for decision."

Capitalist crisis is thus defined as a period of decisive change within the accumulation strategies of the mode of production. For our purposes the definition of "crisis" is not meant to include the negative implications colloquially assigned to the modern use of the term, i.e., the term needs to be understood as being value neutral and devoid of any sense of pathology. For our purposes, a crisis is a turning point, a moment of change, the results of which will benefit some at the expense of others.

Many theorists argue that crisis is an inevitable and inherent component of capitalism, and may indeed be cyclical (e.g., see Habermas, 1975; Kondratieff, 1979; Paynter, 1988; Sweezy, 1942). Paynter (1988) has summarized these theories and suggested their utility for historical archaeology. According to Paynter, cyclical crises can be ordered into three categories of varying duration: (1) business cycles lasting 1–15 years; (2) Kondratieffs, periods of expansion and contraction that last about 50 years; (3) trends seculaires lasting 200–300 years. He identified the period 1790–1845 as a Kondratieff comprised of a period of expansion from 1790 through about 1820 and a period of contraction between 1820 and the mid-1840s. The years prior to 1790 were a period of contraction and the years following 1845 a period of expansion. This theorized cycle coincides with the development and collapse of coffee production in the Blue Mountains of Jamaica. According to this cyclical theory of crisis, Kondratieffs, and this one in particular, were marked by a reorganization of the nature of capitalist work (Gordon et al., 1982; Paynter, 1988:415–17). This certainly was the case in Jamaica, as the 1830s witnessed the abolition of slavery and the introduction of wage labor.

Crisis can produce change in any number of the facets of global capitalism. Depending on the historical moment, it may be more profitable for capitalist elites to manipulate one or more aspects of the relations and/or forces of production, which simultaneously react to and create crisis in other spheres of capitalist social relations. In the case of Jamaica, the abolition of the slave trade and slavery disrupted the social relations of production on which the planters depended. Equally significant, British capitalists began to exploit sugar-producing areas in the Indian Ocean and the mainland of South and Central America. To the Jamaican planter class, the resulting increase in the relative cost of Jamaican sugar production and the simultaneous fall in the price of sugar produced on East Indian estates developed into a crisis. It was this crisis that precipitated the social and spatial changes analyzed in this book.

Crisis and Material Culture

The ability of capitalism to persevere depends on its ability to reorganize itself during and following periods of crisis. Such reorganization does not follow any kind of predetermined or inevitable path;

changes in the culture of capitalism are created by human action for specific historical reasons. The ability of one group of capitalist elites or another to push their specific interests can result in a reformulated capitalism that privileges one sector of the economy over another. Historically, such manipulations involve the struggle for the control — or at least the cooperation or complacency — of the state apparatus that in turn influences the configuration of social relations. To successfully manipulate a given configuration of the capitalist mode of production to maximize profit and wealth accumulation, the elites in question must legitimate and naturalize the changes in social structure that they desire. One way to accomplish this is to change the meaning of material culture.

Historical archaeologists have long believed that the study of material culture, whether recovered from excavations or not, is well within the purview of the field (e.g., Deetz, 1977). In recent years, some historical archaeologists have begun to consider how the relationship between material culture and social change might be manifested. For example, Leone (1995) has argued that federal period architecture constructed in Maryland in the years following the War of American Independence was created to reify the new sociopolitical order that was emerging in the United States. Federal style buildings, including churches, hospitals, the Statehouse, and prisons, "were planned and built as celebrations of democracy" (1995:257). Taken as an assemblage, these buildings explicitly declared a unity with republican Rome and a distance from Baroque notions of hierarchy, as their construction and use was intended to "single out each citizen and . . . also invited each citizen to monitor all the rest" (1995:257). This argument contends that the creation of a disciplined citizenry in the United States required the creation of a self-regulating, self-sustaining population of citizen-individuals. The adoption of neoclassical architectural forms, complete with domes and circular rooms, created not only a link between republican Rome and republican Maryland but reinforced the new political system through the use of panoptic public buildings. In this situation, the new "style" of material culture was intended by its newly empowered creators to be part of a process of legitimizing change within the political structure of British North America. In this example, a group of previously regional elites (the American revolutionaries cum federalists) used material culture to manipulate a manifestation of the crisis of the eighteenth century — the War of American Independence — to legitimate their ascension to the role of core elite.

Leone's argument suggests that elites must have the cooperation of primary producers on whose labor the elites depend. While cooperation can be based on either direct or threatened coercion, more subtle manipulations are more likely to succeed. Leone defines this form of labor control through symbolic manipulations as a process of ideological legitimation. Such subtle manipulations of the working class are less likely to result in violent (and expensive) resistance, and thus would appear to be the rational choice of elites attempting to reorder capitalism. Significantly for historical archaeology, this cognitive process is negotiated through material culture. By using material culture to reinforce changes in the logic of capitalism or the flow of capital, elites can justify change (Delle et al., in press; Leone, 1984, 1988a, 1995).

An effective way of legitimating change in social relations is to change the relationship working people have with material culture, by, for example, introducing new forms of material culture or by making certain commodities more accessible to the working class. Even under the most favorable of circumstances for the elite, however, the responses to newly introduced material culture are often idiosyncratic, or at least highly unpredictable. The meanings that people place on material culture, and the meanings within a given society that material culture elicits and creates, are often arbitrary and/or elicited at the level of the individual; such material changes, then, must be understood as being dialectical. The meanings that the elites who create, market, and distribute the material culture intend to endow, and the social relationships they hope to create (or maintain), clash dialectically with the meanings endowed by the working class. The resulting synthesis creates new social conditions and new cultural meanings whose definition may not be solely controlled by either the manipulators or the manipulated.

For example, David Harvey (1990) has argued that the creation and spread of Henry Ford's system of production ("Fordism") in the first third of the twentieth century can be understood as an attempt to use material culture, and people's relation to material culture, as a solution to the crisis of overproduction that resulted from the mechanization of industrial capitalism that occurred in the late nineteenth century. The purpose behind the 40-hour week/$5 per day assembly line system was not only to regulate production, but also to create a class of consumers capable of purchasing the material culture they produced. As Harvey (1990:125–126) argued "What was special about Ford . . . was his vision, his explicit recognition that mass production meant mass consumption, a new system of the reproduction of labour power, a new politics of labour control and management, a new aesthetics and psychology, in short, a new kind of rationalized, modernist, and populist democratic society." The key to the construction of this new society was the relationship people had to material culture. However, this type of production was resisted in both North America and Europe by labor. Primary producers were hesitant to adopt a system of production that alienated them completely from craft skills and the process of design. This dialectic, between the Fordist capitalists who attempted to reconfigure both production and consumption and the workers who resisted alienation from production and commodity fetishism, resulted in the development of leftist movements in both Europe and the United States (Harvey, 1990).

Material culture, as a social signifier, can be used both to reinforce and to reorganize existing social hierarchies. Daniel Miller (1987) has argued that in periods of intense social stratification, material objects will be used to reflect social hierarchies. Social regulations, like sumptuary laws, can be used to regulate access to certain material items; since access to these items will thus be limited to members of certain social groups or classes, the mere possession of these goods will display and reinforce the social position of the possessor. In times of crisis, which Miller described as periods of breakdown in the relationship between material objects and social status, people lower in the social hierarchy will emulate the material culture of members of the elite in an attempt to "realize their aspirations towards higher status by modifying their behaviour, their dress, and the kinds of goods they purchase" (Miller, 1987:134–136).

When capitalist elites create new forms of material culture with the intention of redefining the conditions of social relations, change can indeed result, but that change can manifest into something entirely unexpected. The manipulation of crisis can theoretically backfire and empower those it was meant to exploit. By emulating or coopting the very material culture meant to suppress them, members of lower social strata can challenge the social structure that the material culture was intended to reinforce. Such was the case in nineteenth-century Jamaica.

"MIDDLE RANGE" THEORY: SOCIAL SPACE AND THE PROCESS OF SPATIALITY

In this study, I consider space to be a class of material culture that can be used to manipulate human behaviors (Lefebvre, 1991:68ff; Leone, 1995; Rowntree and Conkey, 1980). Specifically, I investigate how space was manipulated by colonial agents to elicit specific social responses among the exploited classes in Jamaica during a period of crisis-driven restructuring in capitalism. The central proposition of this project is that the organizational logic of space (e.g., house forms, field arrangements, town plans) is an intricate and inherent element in any such period of labor restructuring. In this section, I examine how a specific type of spatial analysis can be considered as a "middle range" theory that links the material remains of nineteenth-century colonial spatial systems to the general theory of capitalism I have just outlined.

To discuss how the archaeological, documentary, and cartographic remains of Jamaican coffee plantations are linked to the crisis of the early nineteenth century, it is necessary to define space as a form of material culture. The first step here is to define what I mean by "material culture." Although many archaeologists have produced definitions of material culture, the definition I find most adequate is that of Martin Wobst, who defines material culture as a reflexive material product of and precedent to human behavior (Wobst, 1977, 1978, 1989). By this Wobst means that to be considered material culture, an object or thing must both be produced by human behavior and in turn effect the pattern of subsequent human behavior. In defining space as a form of material culture, I thus contend that space is not a natural phenomenon, but is produced or mediated by human behavior to elicit certain behaviors.

A number of archaeologists, philosophers, and cultural geographers agree that while space is created, mediated, or defined by human behavior it also creates, defines or mediates human behavior (Delle, 1996; Lefebvre, 1991; Orser, 1988b; Rowntree and Conkey, 1980; Wobst, 1989). As a precedent or frame to human action, space limits behavior. As Lefebvre (1991:73) stated, "the outcome of past actions . . . space is what permits fresh actions to occur, while suggesting others and prohibiting still others." It is this nature of space as both the created and the creating that informs this study. Specifically, I believe that during periods of crisis, colonial capitalist elites created specific spatial forms that were intended to create specific behaviors and attitudes in the laboring class and to legitimate changes in the colonial social order.

The specific definition of space that I have adopted is based primarily on the work of several geographers and philosophers who have recently sought to redefine the meaning of space as both creation and creator of social relations, i.e., as material culture. By adapting a synthesis of the work of geographers J. B. Harley (1988), David Harvey (1982, 1990), Edward Soja (1989), and Benno Werlen (1993), and the philosophy of Henri Lefebvre (1991), I suggest that space is constructed of three interrelated phenomena: material space, social space, and cognitive space (Werlen, 1993). These different manifestations of space exist simultaneously and are interdependent; one cannot exist in isolation from any other; changes in one will theoretically result in changes in the others (Soja, 1989:120). As an individually experienced phenomenon, space will be understood differently depending on a given individual's perception and frame of reference. For example, in stratified societies different socially constructed and understood spaces will exist in dialectical conflict with alternative spaces based on social position within that society (Lefebvre, 1991:294; Orser, 1988c:324).

To understand how the various spatial phenomena interact, it is necessary to provide definitions of each in turn. Material space is the empirically measurable universe that has been created and/or defined by humans. It comes in an infinite variety of forms, some directly created by human hands, others mediated by human definition. Any room, factory, or plantation is an example of created material space. No one can doubt that any room is created (designed and built) by human beings. Somewhat more ambiguously, what might be superficially defined as pristine "natural" space in, for example, a national park or forest, is really an example of mediated material space, its meaning dependent on the cultural definition of "nature" (Soja, 1989:132). The meaning of the material space will change as the larger social or spatial context changes.

Material space includes what many, including David Harvey (1982:233), define as the "built environment." According to Harvey, the built environment, constructed by humans, is comprised of use values embedded in the physical landscape, which can be utilized for production, exchange, and consumption. Significantly, he sees the built environment as "general preconditions for and direct forces of production" (Harvey, 1982:233). Elements within the built environment, often immovable, continue to exist even after the mode of production under which they were built, and thus the specific use for which they were designed, is no longer in effect. When modes of production change, the social meanings implicit within elements of the built environment, material space, also change. The use values of such elements of material space thus also change (Harvey, 1982:233–234). As specific elements of the built environment become redundant or obsolete, they often fall into ruin, effectively becoming part of the archaeological record.

As a social phenomenon, social space is by nature more ambiguous and more difficult to define than material space. In my usage the term refers to spatial relationships that exist between people and that are experienced in material space. Social space is the complex set of relations that define a person's spatial relationship with other people and with material space. It can be experienced on either the personal or the cultural level. In the contemporary United States, for example, the uncomfortable feeling people experience when somebody stands too close ("invading my space") results from cultural convention and varies depending on the relations of power and intimacy that exist between two people (Hall, 1966:103–112; Orser, 1988b:83). On a larger scale, social space defines how access to material space is allocated to members of any social group and defines appropriate behavior within certain material spaces. In stratified societies, members of the elite classes have the power to define social space by controlling material space. For example, in the American South, plantation slave housing might be specifically located in relation to the fields or the great house (Orser, 1988c:323–324). Such decisions would be made by the planter elite, not the enslaved. In effect, social space is a key variable in the definition of the relations of production that is continually being constructed and negotiated.

The third quality of space relative to this project is what I have chosen to call cognitive space. As the term implies, cognitive space is a mental process by which people interpret social and material spaces. Incorporated in this quality of space is what Lefebvre (1991:40) defines as representing space, the process of cognitively defining and interpreting space. The concept of cognitive space also defines the processes of rendering space as a map, globe, atlas, or a verbal or written description of space; in short, a symbolic representation of the world or part of it (Harley, 1988:58; see also Harley, 1994, and Lefebvre, 1991:39–46). Cognitive space encompasses the idealization of spatial forms. Such idealizations include the imagined or completed blueprints and plats of architects and engineers, a suburbanite's conception of the perfectly landscaped yard, or the design for an Annapolitan formal garden (Leone, 1988:250ff). Cognitive space can be understood as both the conception of social and material spaces that do not yet or may never exist and the interpretation of those spaces that will or do exist.

The concept used to bind the three qualities of a space into a holistic experience of the world is what Soja has defined as "spatiality." In Soja's conception of spatiality, space is simultaneously the product and producer of social relations, somewhat analogous to Wobst's definitions of material product and material precedent (Soja, 1989:120; Wobst, 1978, 1989). Spatiality encompasses all three levels of space as a congruent whole: all three exist simultaneously and together create a material, cognitive, and social place and process. According to Soja (1989:79), spatiality is "the created space of social organization and production." Specific spatialities thus define specific behaviors and social relations. Consequently, material spaces will be designed and created with the intention of introducing specific behaviors. These behaviors in turn are designed to create or perpetuate a set of social relations of production (Foucault, 1979:195–228; Lefebvre, 1991:349; Paynter and McGuire, 1991:15; Tilley, 1990). This concept of spatiality is implicit in Mrozowski's analysis of industrial Lowell in which he argued that the early nineteenth-century layouts and proxemics of mills, workers' housing, and overseers' housing were designed *with the intention* of controlling the behavior of the working class. The spatiality — the interaction between material, cognitive, and social spaces — was designed to define the rules of behavior and thus the terms of the relationship between classes (Mrozowski, 1991).

It must be considered, however, that in stratified societies spatialities will be understood and manipulated differently by members of different social classes (Paynter and McGuire, 1991:11–12). Definitions of space will therefore differ along class, gender, and racial lines. On the one hand, space will be used by elites to create and reproduce conditions of social inequality, while on the other, it will be used by primary producers to resist those conditions (Paynter, 1982:22). For example, Upton (1988) has suggested that eighteenth-century Virginia planters imposed domestic structures on the landscape that differed greatly in form and scale from slave housing. Although such hierarchical landscapes were designed to legitimate an oppressive social hierarchy, the enslaved primary producers endowed that space with meaning distinct from that intended by the elites. In the postbellum United States, southern planters manipulated social space to regulate access to plantation land by former slaves in order to keep the elites "in the exalted position of planter and the freedmen in the unenviable position of dependent farmers" (Orser, 1988b:141).

Spatialities of Colonial Production: Applying This Spatial Model

Soja (1989) has suggested a framework by which the concept of spatiality can be used to analyze change within the capitalist world system. According to his analysis, the continuous expansion and thus the very survival of capitalism as a political economy requires the perpetual redefinition of spatialities. Soja (1989:27) discussed such periods of social and spatial restructuring in terms of "modernization," which he defined as "a continuous process of societal restructuring that is periodically accelerated to produce a significant recomposition of space-time-being." Soja argued that the earliest crisis of global capitalism peaked between the years 1830 and 1851 (see also Wolf, 1982:304). Following these years of crisis there were decades of "explosive capitalist expansion in industrial production, urban growth and international trade, the florescence of a classical, competitive, entrepreneurial regime of capital accumulation and social regulation" (Soja, 1989:27). He might also have added that in this period of restructuring, wage labor replaced slave labor throughout much of the European world system, including Jamaica (Bakan, 1990). The Long Depression of the late nineteenth century, the Great Depression of the 1930s, and the current long crisis in capitalism can be understood as similar periods of crisis-driven restructuring (Soja, 1989:28). The results of these crises have been newly dominant systems of accumulation and oppression, which have required new definitions of space (O'Connor, 1987:92–93; Soja, 1989:28). I consider the situation in early nineteenth-century Jamaica as a case study within Soja's first hypothesized period of global capitalist crisis and restructuring. During this period, from about 1790 through the early 1850s, the political economy of the British world system experienced significant structural change. The regional elites in Jamaica reacted to this set of changes by attempting to redefine certain spatialities, including the production of new spaces of production, an intensification of the commodification of space, and the introduction of new agricultural commodities. The creation of new spaces was made in the attempt to reinforce a capitalist system that privileged white male capitalists over black male producers and all women.

Harvey expounded on this idea of spatial restructuring, which he defined as a "spatial fix," in *The Limits to Capital*. He suggested that crises tend to be experienced on the local level, hence the crisis affecting sugar production in Jamaica need not have affected sugar production in, say, Mauritius, where, in fact, sugar production was on the increase (Deerr, 1950). Harvey argued that the problems within capitalism cannot "be resolved through the instant magic of some 'spatial fix'" (Harvey, 1982:431). However, certain economic crises are "felt at particular places and times and are built into distinctive regional, sectoral and organizational configurations" (Harvey, 1982:431). It thus follows that crisis can be mitigated through switching the flow of capital and labor between sectors and regions or "into a radical reconstruction of physical and social infrastructures" (Harvey, 1982:431).

I hope to demonstrate in the following chapters that during a significant period of crisis within the capitalist system, Jamaican elites attempted to manipulate social relations through the introduction of new material, social, and cognitive spaces. In the case under examination here, the crisis experienced in the sugar-producing sector of the early nineteenth-century Jamaican economy — by far the most important sector — resulted in the reconstitution of the regional economy through the creation of a large coffee sector. The Jamaican planters were able to manipulate the crisis in world coffee production facilitated by the revolution in the French colony of St. Domingue (later the country of Haiti). The uprising in this colony resulted in the collapse of French-controlled agrarian capitalism in St. Domingue, which in the 1760s had been producing nearly as much sugar as all of the British West Indian colonies combined (Watts, 1987:299). Prior to the revolution, St. Domingue was the world's leading producer of both sugar and coffee (Geggus, 1993). The Jamaican planters were able to successfully manipulate the crisis experienced in their own sugar industry by expropriating the knowledge of coffee production from French emigré planters who sought refuge on Jamaica. In doing so, new regions within the island were placed under capitalist agrarian production. In Harvey's terms, the crisis in the sugar sector led to the development of a new sector in coffee. This development resulted in the production of new capitalist spatialities in the previously undeveloped interior mountains of Jamaica. Hence, capital that would otherwise have been invested in intensifying sugar production was diverted to a new crop. The production of that new crop required capital investment, the creation of new productive spaces, and the diversion of labor to newly opened interior lands.

CONCLUSION

The creation of a Jamaican coffee sector was not enough to offset the global crisis of agrarian production. The "spatial fix" that resulted in the creation of new spaces of production in the Jamaican mountains did create profits for a small group of planters, but their prosperity was short-lived. By the 1830s the dialectic between freedom and slavery culminated in the abolition of West Indian slavery. The resulting crisis in the relationship between labor and capital precipitated yet another series of spatial restructurings. In this case, the meaning of land ownership changed. Possession of land by freedmen was constantly challenged by the former slave masters, who themselves were searching for new ways to accumulate capital. In the decades following emancipation, Jamaica experienced radical reconstruction of physical and social infrastructures — the material and social spaces of the island were transformed as the heretofore enslaved working class developed into an agrarian peasantry. Conflict arose between the peasantry and agrarian capital. Many of the emancipated peasants attempted to transform the Jamaican countryside from a space of agrarian capitalism into a space of subsistence production. Many of these attempts were thwarted by the agrarian capitalists, who did not want to see an independent peasantry develop, but strove to create an underemployed agrarian proletariat that was dependent on plantation-based wage labor (Holt, 1992). A conflict in the construction of material and social spaces — or in Harvey's terms "physical and social infrastructures" was the inevitable result of this class struggle to define space in Jamaica.

In considering the historical processes at work during this period of transition within capitalism, this study will examine three interconnected phases of spatial restructuring in the Blue Mountains of Jamaica. The first examines the cognitive processes of planning the initial spatialities of coffee production. The second considers the reconstruction of landscapes into coffee plantations beginning in the 1790s. The third examines the reconstruction of space following emancipation when the laborers and the planters imposed dialectically competing understandings of material and social space on the mountain landscape. To this end, the spatial model I have outlined in this chapter should be considered a form of middle range theory. Using this model, I will suggest how space as material culture was used to negotiate these social changes that transpired in nineteenth-century Jamaica. To begin that analysis, I will first outline the crisis of the nineteenth century by examining the historical context of coffee plantations in early nineteenth-century Jamaica.

3

The Historical Background

The Jamaican Political Economy, 1790–1865

The present times bear a most gloomy aspect truly discouraging to us West Indians. Numbers in this country must be ruined, many are already so; however those who are able to bear up through the impending difficulties will reap the benefits thereafter. The present crisis is past all human calculation. When and what will be the end must be wrapt up in the womb of time, for to hazard conjectures would nowadays be idle in the extreme.
—John Mackeson, Jamaican coffee planter, in a letter to his brother, 1808

INTRODUCTION

During periods of acute social and economic crisis that result in the restructuring of relations between the dominant and the dominated, some factions among the ruling class of a given stratified society will strive to maintain or reestablish the legitimacy of their elevated place within the social order. A key element within the greater process of legitimation is the negotiation of the meaning of material culture (Leone and Potter, 1988; Miller and Tilley, 1984; Shanks and Tilley, 1987); a crucial part of such recurring negotiation is the constant redefinition of space (Lefebvre, 1991:349). In the chapters to follow, I will consider how space was manipulated in nineteenth-century Jamaica, particularly by analyzing manifestations of the three elements of space outlined in Chapter 2. As this is inherently a historical study of plantation life in Jamaica, I will also discuss the historical processes that resulted in the spatial restructuring of nineteenth-century Jamaica. The historical review presented in this chapter considers how contemporary observers perceived the historical crisis gripping Jamaica.

Nineteenth-century observers, modern historians, and various social theorists have argued that the British West Indian colonies were experiencing a phase of crisis during the late eighteenth and early nineteenth centuries (e.g., Bigelow 1970 [1851]; Holt, 1992; Soja, 1989; Williams, 1944). This crisis resulted from structural changes in the global political economy. Edward Soja (1989:27) has defined this period of crisis as ending in 1850; he characterizes this time as being the first of a sequence of crisis-driven periods of "modernization." Crucial to the history of Jamaica, wage labor replaced slave labor throughout the British colonial world (Bakan, 1990; O'Connor, 1984:29).

During the hegemony of mercantile capitalism, rival European metropolitan cores competed for control of colonial peripheral spheres from which they extracted surplus, either through tribute, unequal exchange, or through the exploitation of Native American or African slave labor (Paynter, 1985; Wallerstein, 1980; Wolf, 1982). This exploited labor was used to produce commodities, usually in the form of unfinished raw materials or addictive agricultural products, that were exchanged in the European commodities markets (Mintz, 1985a). The crisis of the early nineteenth century resulted in the florescence of competitive capitalism, as the dominant social relations of production in the colonies were restructured to conform to a regime of wage labor rather than slavery (Bakan, 1990). There was an accompanying shift in temporal and spatial structuring to facilitate this wage-based labor exploitation (Foucault, 1979; Paynter and McGuire, 1991). People were systematically alienated from production to the point where labor power was defined by time: wages were paid by the hour or day or week, slices of time worked, rather than by the value of the product of that labor. Efficiency became the measure of capitalist competition: Whoever controlled the time and cost of production most effectively became the dominant capitalist. Although there is no definitive explanation as to why and how capitalist crises occur, there is general agreement that crisis does periodically occur and that such times are periods of opportunity as well as despair and ruin. Historical crisis in one sector of the world economy creates opportunities for various rival sectors to arise and challenge existing hegemony in the capitalist order (e.g., Harrison and Bluestone, 1988:3–20; O'Connor, 1984, 1987:14–48; Paynter, 1988:418).

Such was the case in Jamaica. Under the mercantile phase of capitalism, the Jamaican political economy was dominated by the sugar sector. However, increased consumption in Europe, the availability of cheaper sugar produced in other colonial regions, the disruption of commodity flow during the North American wars in the late eighteenth century and the Napoleonic Wars in the early nineteenth century, all combined to drive the Jamaican sugar industry into crisis. Although sugar production continued to increase, reaching its recorded peak of nearly 90,000 tons in 1820 (*Jamaica Almanac*, 1846), Jamaican sugar planters were unable to maintain an increasing rate of profit. As production increased, the price on the London market decreased. Competition with other sugar-producing colonies, restructuring of the labor system, and the eventual establishment of free trade as the official policy of the British government resulted in hard times for the Jamaican sugar industry. Between 1830 and 1840, the period during which slavery was abolished, Jamaican sugar production decreased by nearly two-thirds. By this time, and under a new system of wage labor, the Jamaican planters could not compete in the expanding global sugar market, and many eventually came to ruin (Higman, 1988; Mintz, 1979). The Jamaican sugar crisis, which began as early as the 1790s, provided the opportunity for a new commodity sector to take root in Jamaica. The crisis of the sugar planters became an opportunity for smaller investors to create a new sector in colonial agricultural production in Jamaica: coffee.

It is generally believed that the capitalist world economy was experiencing a crisis phase as the eighteenth century came to a close (Mintz, 1985a; O'Connor, 1984; Paynter, 1988; Soja, 1989; Wallerstein, 1989; Wolf, 1982). It was during this crisis that Jamaican planters began to experiment with coffee production. Although the planters invested heavily in this new commodity, many of the same structural and historically contingent phenomena that precipitated the decline of the sugar industry eventually caught up with the coffee planters. It was during this crisis that the structural logic of capitalism shifted from the mercantile to the competitive phase. Because this transition in the global economy precipitated the introduction of large-scale coffee production in Jamaica, and the resulting reconfiguration of space into coffee plantations, I believe that it warrants some further discussion.

CRISIS IN THE EUROPEAN SYSTEM: 1763–1834

The collapse of mercantile capitalism as the dominant mode of production in global capitalism can be explicated by examining a series of political and economic developments that affected the European world economy in the years from 1763 to 1834 (O'Connor, 1984; Paynter, 1988; Soja, 1989). Over the course of this period, the core elites of the major European powers lost political control of many of their respective mainland American colonies (Wallerstein, 1989:193). Simultaneously, the emergence of competitive capitalism created a new order among European elites whose agenda at times conflicted with the interests of the West Indian elites who had profited most during the mercantile era (Bakan, 1990; Engerman and Eltis, 1980; Walvin, 1980; Williams, 1944).

There can be no doubt that the fate of every one of the British colonies in the Western Hemisphere was influenced by incidents occurring on mainland North America. Prior to the War of American Independence, for example, many of the British Caribbean colonies, including Jamaica, depended on trade with North America to obtain foodstuffs and livestock (Sheridan, 1974:106–107). A number of historians have traced the beginning of the crisis of the nineteenth century to the North American experience in the Seven Years War, a struggle that the British won and the French lost (1756–1763). A significant result of the war was the termination of French sovereignty over the northern tier of North America, or more precisely, the seizure of political and economic power from the French by regional elites with political and economic ties to the English core in London (Wallerstein, 1989). Braudel has suggested that the conclusion of the Seven Years War led to the development of animosity between core and regional elites in the British world, as the core elites attempted to extract additional surplus from the American regional elites to finance war debts (Braudel, 1984:412; see also F. Jennings, 1988). Wallerstein has argued that during the post-Seven Years' War era the regional elites organized resistance to the increased extraction of resources (Wallerstein, 1989:214). This resistance was violent at times and developed into armed rebellion against the core elite and those among the regional elite who chose to comply with the added demands rather than risk treason (Braudel, 1984:401–409; Wallerstein, 1989). This animosity eventually led to the War of American Independence, which, of course, resulted in the termination of British sovereignty over most of the eastern seaboard of the North American mainland (Wallerstein, 1989:207–211). With the British recognition of the independence of the United States, a regional elite, or more precisely a group of regional elites, created the infrastructure for the development of a new center of political economic power in the Western Hemisphere (Braudel, 1984:410–411).

The ramifications of the newly developing cores in North America were multifarious. Significantly, the development of a constitutional republic, albeit controlled by a small group of urban and landed elites, encouraged radical intellectuals in France to steer their own social crisis of the 1780s into a republican revolution (Lefebvre, 1947:21, 155; Schama, 1989:24; Wright, 1995:41–43). This internal national crisis led to another series of wars, fought not only on the European continent, but in the West Indian colonies as well. One result of the wars, which persevered until 1815, was Napoleon's Continental System, which closed continental ports to British, and consequently British West Indian, imports (Wallerstein, 1989:117–118). Wartime embargoes and shortages led to temporary prosperity for some sugar producers, as scarcities drove prices up; however, the disruption of trade as a result of maritime warfare created a siege mentality in Jamaica, adding fuel to the fire of crisis (Wright, 1966). A critical turn in the late eighteenth-century wars, the Haitian Revolution, precipitated the collapse of French coffee production on the colony of St. Domingue, and directly resulted in the intensification of coffee production in Jamaica.

During the French Revolution and the Napoleonic Wars, political and social crises were coming to a head in the French Antilles, notably in the sugar colony of St. Domingue. The domestic upheavals in France were not lost on the colonial populations, enslaved or free. The social crisis extended to the Caribbean, fired by the ideology of freedom and equality articulated if not practiced by the American and French revolutionary elites (Blackburn, 1988; James, 1963). The slave population of St. Domingue rebelled, eventually establishing the Empire of Haiti, the first independent nation controlled by members of the African diaspora (James, 1963; Thompson, 1987:301–355). During the course of the Haitian revolution, some of the planter elites fled to neighboring islands, notably Jamaica, and attempted to regain their lost place in the colonial hierarchy (James, 1963; Thompson, 1987:316; Williams, 1970).

The social and political events of the late eighteenth century resulted in extensive economic restructuring of the European world economy during the nineteenth century (O'Connor, 1984; Soja, 1989; Wallerstein, 1989). Unlike mercantile capitalism, competitive capitalism was characterized by factory production of commodities; the restructuring of social relations within the industrial base eroded the system of social relations that had existed in the European-controlled world under mercantile capitalism. As an increasing number of common lands were enclosed, a vast number of British workers migrated to the city, developing into an urban proletariat selling their labor power to factory owners for wages (Johnson, 1996). With the growth of this urban proletariat came new pressures to create a source of caloric energy cheap enough to prohibit mass starvation, if not malnutrition, in the growing industrial cores. According to the argument championed by Sidney Mintz (1979, 1985a,b), that source became increasingly sugar based. Given the expanding European sugar market, one might logically assume that increasing demand would result in increasing production and sales of Jamaican sugar. Unfortunately for the sugar producers, this was not the case. In the early nineteenth century, the Jamaican sugar industry experienced a series of setbacks which aggravated the already depressed situation. As a result of Napoleon's continental embargo of British shipping and the disruption of the flow of sugar from St. Domingue because of the Haitian revolution, several European governments, including Napoleon's, encouraged the development of European beet sugar production, which decreased the European demand for West Indian sugar (Deerr, 1950:471–500). Furthermore, in part resulting from the loss of political control over the North American mainland and Haiti, English, Dutch, and French core elites sought to exploit newly constructed sugar plantations in the East Indies and the Indian Ocean (Braudel, 1984:485–496), further jeopardizing the success of the West Indian sugar industry.

A third development that affected the ability of the West Indians to compete in the global market was the progressive replacement of slave labor with wage labor in the European sphere of influence. Although opposed by the West Indian sugar interest, Parliament passed legislation that abolished the African slave trade in 1807. Some years later, in 1834, slavery was legally abolished throughout the British Colonies (Bakan, 1990; Butler, 1995; Green, 1976; L. Jennings, 1988). Accounts written by nineteenth-century travelers to Jamaica described a depressed situation on the island, which the planters believed to be caused by the loss of capital they had suffered as a result of emancipation, as well as the loss of tariff protection for West Indian sugar. One traveler put it this way: "The present ruinous condition of Jamaica is ascribed by its inhabitants mainly to three causes, the abolition of slavery in 1834, the inadequate compensation paid to the owners of slaves, and the repeal of the protective duty on British colonial sugar" (Bigelow, 1970 [1851]:71).

The crisis for the Jamaican sugar planters is reflected in the falling price for their sugar on the London market. According to Noel Deerr (1950), the price for Jamaican sugar on the commodities exchange consistently fell from its 1795 peak of 80 shillings (£4) per hundredweight to a low of 24 shillings (£1 4s) per hundredweight in 1830 (see Figure 1). As prices fell, sugar planters attempted to maintain their profits by increasing overall production. While the crop of 1790 was 55,600 tons, by 1805, in reaction to the falling price, production nearly doubled to a crop of 99,600 tons. By 1810, the overproduction of

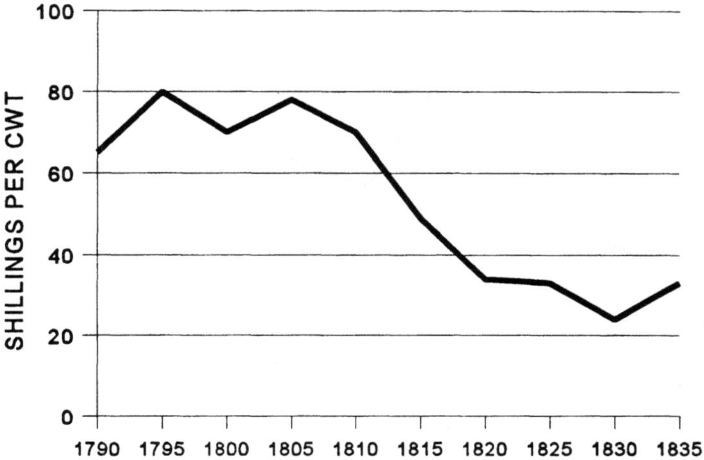

Figure 1. London price for Jamaican sugar, 1790–1835. Source: Deerr (1950).

sugar resulted in a price collapse; in these 5 years the price dropped from 70 shillings to 49 shillings per hundredweight. By 1820, the price had dropped to 34 shillings, less than one-half the price received only 10 years earlier. Low prices resulted in a collapse in production, as continuously smaller crops were reported (Deerr, 1950; see Figure 2, this volume). The rapid devaluation of sugar as an agricultural commodity coupled with the availability of sugar from Europe and colonies in the Indian Ocean eroded the economic power once wielded by the West Indian sugar interest.

As social and economic restructuring began to affect the price of West Indian sugar, planters began to experiment with alternative crops. The most notable of these was coffee. Large-scale Caribbean coffee production began in the French colonies, probably in Martinique (Watts, 1987:503). The French colonies of Martinique, Dominica (prior to cession to Britain), St. Domingue, and Guadeloupe, the Spanish colonies in Cuba and Puerto Rico, and the British colonies of Montserrat and Jamaica all produced coffee by the early nineteenth century (Watts, 1987:503). However, in all of these areas — with the exception of Dominica — coffee remained secondary in importance to sugar (Berlin and Morgan, 1993:332). Incidentally, Geggus points to 1763 — the year the Seven Years' War ended — as the point after which intensive coffee production began in St. Domingue (Geggus, 1993). Prior to 1790, St. Domingue was the world's leading producer of both sugar and coffee (Blackburn, 1988:163; Geggus, 1993). With the advent of the Haitian

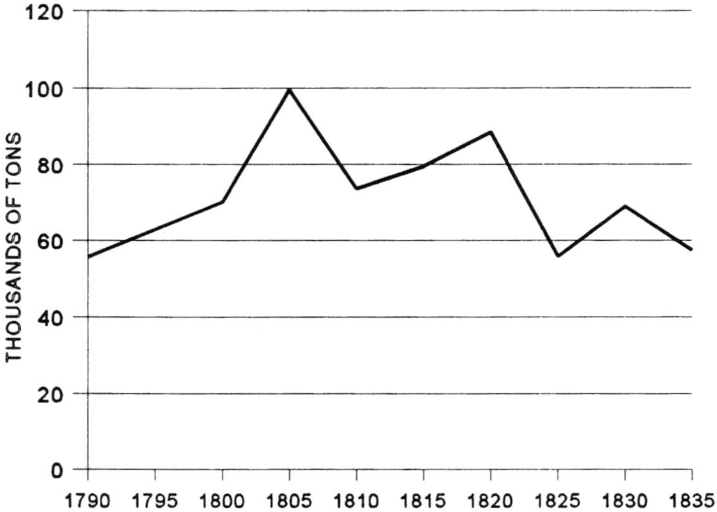

Figure 2. Estimated total production of Jamaican sugar, 1790–1835. Source: Deerr (1950).

revolution in 1790, the production and export of both commodities fell calamitously. A number of French planters sought refuge on Jamaica (Thompson, 1987:316; Wright, 1966); several of these assisted Jamaican planters with the initiation of coffee production. Prior to 1790, virtually no coffee was exported from Jamaica. In 1810, more than 25 million pounds were exported (*Jamaica Almanac*, 1790, 1846).

THE CRISIS INTENSIFIES: LAND AND LABOR, 1834–1865

Despite the mass marketing of sugar to the emerging European proletariat, as early as the 1790s, and certainly through the first half of the nineteenth century, the West Indian sugar interests were faced with crisis and potential ruin. On world markets, they could not compete with sugar produced more cheaply on the fresher soils of Cuba, Brazil, and Mauritius. Following emancipation, Jamaican planters could not control labor from abandoning export commodity production, nor could they produce sugar as cheaply as their competitors in India and the East Indies could with indentured contract labor. Finally, as world sugar production increased, the Jamaican sugar interests suffered from overproduction crises, which Williams (1944:152) has suggested triggered abolition of the slave trade in 1807 and slavery in 1834. In Jamaica, as elsewhere in the Caribbean, these changes in the British political economy resulted in significant changes in the social structure.

Perhaps the most obvious and significant result of the global shift in capitalism that occurred in the nineteenth century was the abolition of slavery in the European West Indian colonies. The termination of Caribbean slavery was a drawn-out process that began with the uprisings in Haiti in the late eighteenth century, but did not end until almost 100 years later when slavery ended in the United States, Brazil, and Cuba. In the British West Indies, the abolition of slavery was accomplished in three phases. The first phase was the abolition of the African slave trade in 1807; the second phase was the replacement of slavery with a system of contract bondage known as "apprenticeship" in 1834; the final phase was full emancipation in 1838. As Williams has argued, the final nail in the coffin of the West Indian plantation economy was the termination of preferential sugar duties on West Indian sugar, which occurred in 1846 (Bakan, 1990:27; Butler, 1995; Williams, 1944:136).

The competitive capitalist mode of production precludes the use of slave labor, relying instead on free wage labor. In the years between 1834, when the British Parliament enacted the Emancipation Act, and 1838, when the system of "apprenticeship" was terminated in Jamaica, the slave population was transformed into a protopeasantry (Mintz, 1961, 1985b). A number of the large planters who had all but dominated land ownership began to sell or rent land to the peasantry, in the hope of creating a wage labor force to work the plantations (Holt, 1992; Watts, 1987:508). However, the dramatic fall in the price of sugar and the attendant decrease in wages made the legal acquisition of land difficult for the peasant class. The mortgage difficulties faced by the planters as a result of falling prices in London made it increasingly difficult for them to sell their land (Watts, 1987). In addition, the vast quantity of undeveloped land in the interior of Jamaica enticed large numbers of people to carve out their own farms, without the formality of land transfers (Holt, 1992). By the 1860s, the population was predominately a landless peasantry, squatting on Crown land or abandoned estates (Bakan, 1990; Higman, 1988; Satchell, 1990; Watts, 1987).

An additional factor in the equation of crisis for Jamaican sugar producers was the leveling of duties, or import tariffs, on sugar produced in other regions of the world. Prior to 1826, the West Indian sugar interest enjoyed protective duties in the United Kingdom, which equalized the price of cheaper sugar imported from other colonies. In 1826, Parliament passed an act removing the protective duty from sugar imported from British colonies outside the West Indies, i.e., Mauritius and the British East Indies. In 1846, the Sugar Duties Act removed the protective duty from all foreign sugars. From this point on, the Jamaican planters were required to compete on an equal footing with producers from the East Indies as well as those from Brazil and Cuba who were still using enslaved labor (Hall, 1959:83–84).

An important debate within the historiography of the Caribbean and of Jamaica concerns both how the abolition of slavery occurred and was experienced, and why this centuries-old labor system was abolished. Historians considering this issue have generally fallen into one of two camps: (1) those who believe that the abolition of slavery was part of a larger project of liberal reform emanating out of the newly empowered bourgeoisie, who felt slavery a moral evil that had to be abolished, and (2) those who argue that the abolition of slavery was based primarily on changes in the political economy that rendered slavery an obsolete and contradictory institution. The last half-century has witnessed considerable debate between scholars subscribing to these two schools of thought. Nearly two decades after his death, Eric Williams remains perhaps the most influential of Caribbean scholars; indeed it may be said that it was he who opened this debate half a century ago. In 1944, Williams published his classic work *Capitalism and Slavery*. In this treatise, Williams argued that slavery was necessary for the development of agricultural capitalism in the West Indian and subtropical mainland colonies of the New World. Moreover, he argued that the rise of industrial capitalism with its wage-based labor system led directly to the decline of slave-based plantation agriculture. Williams's correlation of abolition with the concurrent fall of mercantile capitalism and the rise of industrial or competitive capitalism has become known as the "decline" thesis (Drescher, 1977:7ff).

Williams's thesis, while elegant, has by no means been universally accepted. A number of notable scholars interpret the abolition of slavery as a consequence of a changing moral ideology in Great Britain, quite independent from economic developments (e.g., Drescher, 1977). Of particular note is Roger Anstey (1968, 1975, 1980), who interpreted the abolition of first the slave trade and later slavery itself as part of a process of liberal reform with roots in the eighteenth-century Enlightenment. Anstey argued that the abolition movement was more ideologically than materially based. He contended that in studying the process of abolition, one must consider that in the eighteenth century "the emphasis on benevolence and the invoking of the principles of nature and utility were themselves marks of a growing disposition to effect change in the area of natural, civil, and political liberty, by legislative action" (Anstey, 1980:20). Thus, the British abolitionists were eventually successful because the *zeitgeist* in which they were operating emphasized the importance of the individual and personal freedom, and also because Parliament was in the position to pass legislative acts that could dictate societal change. Although Anstey believed that the abolition of slavery was more a social than an economic process, he did not deny that the West Indian economy was at a point of crisis in the decades leading to emancipation (Anstey, 1980; Drescher, 1977:6). He does suggest that even though declining in importance, the West Indian political interest was able to delay the coming of abolition: "The West Indians in Parliament . . . were more effective in checking anti-slavery in 1823–33 than twenty-five years earlier" (Anstey, 1980:30).

Seymour Drescher's critique of Williams's thesis is articulated in his 1977 volume, *Econocide: British Slavery in the Era of Abolition*. In this book, Drescher argued that Williams was qualitatively wrong in his assessment of the decline in importance of the West Indian colonies to the British political economy. According to Drescher's calculations, the process of crisis did not begin until after 1807, the year that the African slave trade was abolished by an act of Parliament (Drescher, 1977:20). Because he contended that the value of the West Indies was not declining prior to the abolition of the slave trade, Drescher (1977:184) argued that economic explanations for the abolition of slavery are not sufficient to explain why Parliament would intentionally disassemble a profitable economic system. In his estimation, the abolition of slavery was dependent on the mass mobilization of public opinion in Britain. Drescher (1977:186) concluded that in order to understand why abolition occurred, political, economic, and ideological forces needed to be analytically separated, for in his opinion it "is only by separating the elements that we sense just how they worked for or against each other."

In recent years, James Walvin (1994) has reaffirmed the argument that abolition was brought about by a lengthy political campaign, which was based on the ideological tenet of individual liberty. He rightfully acknowledged that, particularly in Jamaica, the agency of the slaves themselves contributed in no small part to the final abolition of slavery (Walvin, 1994:307). Such organized resistance to the Jamaican slave regime was most dramatically expressed in the great uprising of 1831, which has become known alternatively as the Christmas Rebellion and the "Baptist War" — the latter as the insurrection was blamed on Baptist missionaries who lived among the enslaved population. This was by far the largest slave uprising experienced in the British colonies. Blackburn estimates that between 20,000 and 30,000 slaves were involved in the rebellion, which began on Christmas Day 1831. A well-armed British militia required 2 weeks to suppress the rebellion. At its end, 14 whites had been killed, and property valued in excess of £1,000,000 had been destroyed. In retribution, some 200 rebels were killed in action; another 312 were executed, many on no evidence beyond the testimony of their overseer. Walvin argued that the violence of this uprising convinced many in England that slavery was an institution that could no longer be tolerated (see also Blackburn, 1988:432–433; Taylor, 1885:123).

Thomas Holt has argued that the violent reaction on the part of the planters against both the rebels and the Baptist and Methodist missionaries who were accused of fomenting the rebellion mobilized support for abolition in Britain (Holt, 1992:17). He suggested that the pogrom launched against the nonconformist missionaries in the wake of the rebellion broke an uneasy compromise between the missionaries and the planters. The nonconformists reacted politically, organizing petition drives and utilizing the opportunities opened by the reform of Parliament to elect proabolition members of Parliament in unprecedented numbers. A year after the Baptist War, Parliament was actively studying ways to implement abolition. Hence, the action of the slaves in violently resisting slavery, coupled with their alliances with politically astute nonconformist sects, accelerated the process of abolition (Holt, 1992:17).

These historians, varying in their assessment of Williams's thesis, do agree that the mode of production that had prevailed in the British West Indies was suffering through a period of crisis, turmoil, and restructuring. The predominance of West Indian sugar production in the imperial economy had faded by the 1820s and continued to decline as the British economy expanded. The newly emerging version of the capitalist mode of production emphasized what have become nearly clichéd tenets of competitive industrial capitalism: the importance of the individual, competition, and free trade. The monopolies and price controls that the West Indian planters had enjoyed for decades had come to an end, while an increasingly sophisticated and militant enslaved population grew increasingly intolerant of slavery and began to develop a class consciousness as demonstrated by the large-scale rebellion of 1831. For the planters this crisis led to the dismantling of the slave labor system on which they depended, declining prices for their commodities, and the loss of political power in Parliament.

CONTEMPORARY ACCOUNTS OF CRISIS IN JAMAICA

Although modern historians and social theorists have analyzed the nature of the crisis gripping the Atlantic world during this period, their perceptions are mediated by the passage of time and the process of interpretation. To examine the perceptions of those who actually lived these experiences, I will now consider how contemporary observers interpreted the events of the early nineteenth century. As is often the case with documentary evidence, it is only the thoughts of the elites that have been preserved in historical accounts. Just what the enslaved — and later emancipated — people thought of these experiences is perhaps impossible to reconstruct.

Economic statistics, which were meticulously kept by the government, were available to the contemporaries of the period. Based on anecdotal observation as well as quantitatively demonstrable economic contraction, observers and critics of the West Indian colonies perceived the slave colonies to be in a period of economic decline or crisis (Drescher, 1977:10). Many such arguments were published during the time period under consideration. A number of these observers proposed schemes by which they thought the crisis could be ameliorated, particularly through the reconstruction of social relations. The remarks of three plantation theorists who believed their ideas would have a direct impact on the Jamaican social landscape are of particular note: Edward Long (1970 [1774]), Dr. David Collins (1971 [1811]), and Thomas Roughley (1823). Each spent considerable time in the West Indies; all but Collins resided in Jamaica. All three perceived a crisis gripping the West Indies, and suggested how to manipulate class relations in order to stabilize what they perceived to be an unstable social situation. Significantly, they understood that the social manipulations they suggested needed to be reinforced and reified by the manipulation of space.

In 1774, Edward Long published a voluminous tract titled *The History of Jamaica*. More than a simple historical narrative, Long's three-volume work described the topographic, political, and social situations of late eighteenth-century Jamaica. As a member of the Jamaican planter class, he was significantly biased toward the interests of the planter elites. Born in Britain in 1734, the second son of a Jamaican planter, Long was a direct descendant of Samuel Long, the first Speaker of the Jamaican House of Assembly (the island's legislative lower house), the son of another Samuel Long who was a member of the Jamaican Council (the effective upper house), and the brother-in-law of Sir Henry Moore, a Lieutenant Governor of Jamaica. Educated in Britain, Long moved to Jamaica at the age of 23, on the death of his father in 1757. Through his inheritance and a well-placed marriage in 1758, Long became a man of considerable wealth. After an appointment as Justice of the Vice-Admiralty Court of Jamaica, in 1761, at the age of 27, Long was elected a member of the Jamaican House of Assembly. He was elected Speaker of that body in September 1768; unfortunately for him, the Assembly was dissolved 10 days after his election. In January 1769, because of deteriorating health, Long returned to England, remaining there until his death in 1813.

Long recognized that late eighteenth-century Jamaica was experiencing a crisis in the social relations of production, which he believed stemmed from the racial composition of the island's population. Chief among Long's concerns was the increase in the ratio of African slaves to white inhabitants of Jamaica. He believed that the growing gulf between the number of whites and that of blacks was caused by a diminution in the number of European indentured servants who chose to remain in Jamaica following the term of their servitude. Long believed that given the reality of slavery, the absence of freed indentured servants served to create an unworkable class and racial structure in which the relative number of planters would constantly decrease, the relative number of slaves would constantly increase, and no middle class would evolve. The absence of a European middling class resulted from the inability of former indentured servants to gain access to productive land, which was hoarded by the wealthy planters, as well as from the preference of African slave labor over European indentured labor (Long, 1970 [1774]:386).

Long argued that the expansion of large plantations resulted in marked class stratification of the island. While the gross number of plantations was increasing, the wealthier plantations were too freely absorbing smaller plantations. In Long's mind, such land speculation resulted in a dangerous situation in which a select few white settlers controlled access to production through land monopolies. This in turn led to a decreasing population of whites (Long, 1970 [1774]:386). Furthermore, absentee landowners were draining Jamaica of capital that could be used to support a white working class, and thus contributed to the declining European population:

> The emigration of many owners of property, who of late years have flocked to Britain and North America . . . drained those incomes from the island which formerly used to be spent there in subsisting various artificers, shop-keepers, and other inhabitants, forms the further cause of a very great diminution. (Long, 1970 [1774]:386–387)

Long observed that tensions were growing between the regional elites and the core elites back in Britain. As Long was a member of the planter elite, it comes as little surprise that he blamed what he perceived to be a demographic crisis on the core elites. According to Long, in the late eighteenth century the Jamaica Assembly passed a tax bill that would impose a heavier tax on absentees than on resident planters. Long argued that the local elites had a more secure knowledge of what was best for the colony than did the absentee interests in Britain. He bemoaned the growing influence the absentees had in London, as exemplified by Parliament's eventual overturning of this tax measure that he had hoped would discourage absenteeism (Long, 1970 [1774]:388).

There is no doubt that Long's racism affected the tone of his arguments; he did indeed believe that Africans and Europeans were members of different species (Trigger, 1989:112). His arguments, which superficially could appear to be based solely on his own virulent racism, however, are more complex than simple racist diatribe. Although he did believe that a larger white population would mitigate the possibility of large-scale slave insurrections, he was concerned with the construction of a new class structure that would propel Jamaica and Great Britain into hegemonic control of the global sugar trade. To do so, he suggested that the British abandon the scattered West Indian colonies in favor of concentrated production on Jamaica (by far the most productive of the British sugar colonies in the late eighteenth century). In classical economic terms, his argument was a rational one (Long, 1970 [1774]:402).

Long believed that British planters in Jamaica were playing an increasingly secondary role in the European sugar trade. He proposed a scheme by which he felt the British could overcome their chief rivals, the French planters of St. Domingue, and thus dominate the European market. This scheme called for Britain to abandon the remaining nine West Indian sugar colonies in order to transplant their populations to Jamaica, and thus to promote the "full" cultivation of Jamaica. Long suggested that such a rationalization of production coupled with the elimination of the cost of defending the smaller sugar islands would allow the British to overtake the French, eventually allowing Britain to undersell the French in world markets. The success of this scheme, according to Long, would be the establishment of a class structure based not only on slave labor, but also on the development of a European middle class in Jamaica. He believed the French:

> have pushed their interests in the West Indies, not by fewer taxes, the lower price of Negroes, or the greater cheapness of provisions, and implements of husbandry; but by their ability to furnish double the number of European hands, and by wiser internal regulations. (Long, 1970 [1774]:403)

By "wiser internal regulations" Long meant to imply that the affairs of Jamaica, and thus the profitability of the sugar trade, were dependent on increasing the power of the resident planters to direct the internal affairs of the island.

Among the class tensions perceived by Long was a controversy surrounding vacant land in the interior of Jamaica. He asserted that several hundred thousand acres of interior land had been patented but never developed by the Jamaican planter oligarchy, who held the land in some kind of speculative trust. The monopolization of such large tracts of land, argued Long, prohibited the rational improvement of Jamaica. To illustrate his argument, he claimed that in the mid-eighteenth century, in the parish of St. James, 106,352 acres were owned by a mere 132 people (Long, 1970 [1774]:406–407). In order to relieve such disparity, wherein a handful of wealthy planters could hold land without transforming it into productive spaces, Long supported the implementation of a quitrent. Such a measure would force, by law, the Jamaican planters to declare the extent of their holdings and to pay a tax on unimproved land. He congratulated the Assembly for passing such an act in 1768 (the year in which he was elected speaker!):

> This measure...reflected great honour upon the legislature that passed the act; because: it has generally, and with good reason, been conjectured, that the members of the legislature, being men of large landed property in the island, and some of them unconscionable monopolists, considered the quitrent as a species of land tax, and combined together to excuse themselves from paying it, or to obstruct the making of a public discovery of the large uncultivated tracts in their possession lying useless to themselves, and unbeneficial to the colony or nation. (Long, 1970 [1774]:406)

This passage reveals the tension that existed between the landed oligarchy, and the more "progressive" members of the assembly, like Long, who sought to break the collusion of the larger landholders and to open the Jamaican interior to white settlement.

While Long was concerned with a crisis in the class structure of Jamaica, Dr. David Collins, whose ideas would affect the discourse of space in Jamaica, was concerned with the crisis of social reproduction. Collins, a physician who resided on the island of St. Vincent, originally published his major work, *Practical Rules for the Management and Medical Treatment of Negro Slaves*, anonymously. In the first edition of this treatise, he described himself as professionally educated, as having lived in the West Indies in excess of 20 years, and has having been a self-made planter, who had purchased rather than inherited his enslaved population. He argued that as a result of his own personal investment he took a supreme interest in the health of his slaves, as their ability to work directly affected his ability to produce wealth (Collins, 1971 [1811]:8–9). Collins was described by his contemporaries as both an eminent medical practitioner and a successful sugar planter who created a large fortune for himself through his managerial skills (Sheridan, 1985:32).

Collins's book was published in 1803 and later reprinted in 1811. This was a well-known work in its day, cited by both critics of slavery like the abolitionist James Stephen (1969 [1824]) and by apologists for slavery, like the Jamaican planter Alexander Barclay (1969 [1828]). In this book, Collins delineated what he believed to be the best system of maintaining a population of slaves in the West Indies. As his primary focus was the health of the enslaved population, he explored such topics as pregnancy, diet, clothing, labor discipline, diseases, and the proper design of slave houses and villages.

Collins's work was written at a time when there was great concern among the planter classes about the reproduction of the enslaved populations. As Barry Higman (1976:102) has demonstrated, in the decades prior to the abolition of slavery in 1834, the African population of Jamaica was in a constant state of natural decline; until the African slave trade was abolished in 1807, the planters relied on the continuous importation of slaves to reproduce their labor force. Collins (1971 [1811]:130) was well aware of the demographics that resulted in a natural decrease in population and forced the planters into a dependence on the African slave trade. He believed that there was a heavy economic cost involved in the reproduction of labor through purchase. Collins reported that in the years prior to 1803 it was common knowledge among planters that it was cheaper to purchase adult Africans than to invest in the rearing of children. However, because of the inflation in the price of African slaves, he claimed that in the early nineteenth century, planters needed to rethink their strategy for the reproduction of labor, particularly considering the imminent abolition of the slave trade (Collins, 1971 [1811]:131). Not to improve the health and welfare of the enslaved population would eventually result in crises not only for the African population, but also in the viability of the plantations that depended on these people for labor.

Collins ascribed the population decline primarily to "destructive occupations in an unhealthy climate," but added four corollary causes to explain the negative demographic trend: (1) the small ratio of women brought from Africa,(2) the sterility of African women, (3) the frequency of abortions practiced among enslaved women, and (4) high infant mortality. Interestingly, he went into far more detail in describing these four factors, which he believed could be efficiently remedied, than he did the "destructive occupations in an unhealthy climate." Collins laid the blame for population decrease on the enslaved women, in effect, blaming the victims for the demographic crisis.

Another cause of demographic instability was the overworking of weaker individuals, which resulted from their being expected to work alongside stronger people of similar age. Collins (1971 [1811]:151) recommended that an age-based division of labor be refined to segregate stronger individuals from weaker ones, requiring the more robust to perform heavier tasks than the weak, regardless of gender. As a reward for the harder labor expected of this group, the "strong gang," Collins (1971 [1811]:152) suggested that the planter should institute a task, or piecework system, allowing the gang to quit work when a specific job was finished, a privilege that should be limited to this strong gang.

Collins recommended that those not deemed fit enough to work in the strong gang should be assembled into several other gangs, depending on their ability to work. A second gang, including weaker but fit adults and adolescents, would work at tasks not quite as demanding as those assigned to the strong gang. A "small gang" would consist of younger children and those recently dismissed from the hospital; this gang would be responsible for weeding crops and spreading manure. Children aged 5 or 6 would be assigned to a "grass gang." Under the supervision of an elderly woman, this gang of children would pick grass to be fed to the livestock. Collins suggested that any sickly people be added to the grass gang, so that the planter could monitor their progress back to health and eventually to more productive labor. As for the children, Collins suggested:

> It is of great advantage to introduce your young people early into habits of labour, as so much of it will be required from them in their future progress through life. (Collins, 1971 [1811]:156)

Thomas Roughley, a Jamaican resident planter, was concerned with what might be defined as a crisis of production. Roughley published his influential theoretical tract on the proper management of West Indian estates, *The Jamaica Planter's Guide*, in 1823. In this volume, Roughley described what he believed to be a crisis enveloping the economy of Jamaica. The self-defined purpose of his book was to first expose the roots of this crisis and to offer a rationalized system of plantation management to correct what was wrong. His core idea held that the decreasing exchange value of Jamaican produce, coupled with the increasing costs of production, required planters to reconsider the conventional management strategies then in use (Roughley, 1823:v).

Roughley attributed much of the production crisis to the intentionally inept management of estates belonging to absentee proprietors. According to Roughley's calculation, as many as seven-eighths of Jamaican proprietors were absentees (Roughley, 1823). As absentees, these proprietors would hire an agent (or "attorney") to manage the affairs of the estates. The attorneys, who often managed numerous estates for numerous proprietors, were responsible not only for conducting the commercial affairs of the estates (e.g., shipping produce, selling produce on consignment, dealing with mortgages), but also for supervising the overseers of the estates, and relaying any crucial information concerning the estate back to the proprietor. In return for managing the estates, the attorneys were paid a percentage of the estates' production.

Roughley claimed that these attorneys were often unscrupulous speculators who would often intentionally bankrupt the estates under their supervision. As many of these attorneys would be hired by the merchants who held the mortgages on the estates, as well as the proprietors who took out the loans, they were in the unique position to assist the merchants in acquiring estates, at the expense of the proprietors. Roughley accused the attorneys of scheming cultivation plans with the intention of ruining crops, so that the proprietors would be unable to meet the requirements of their mortgages, which were often secured against future production. As Roughley put it:

> the fairest prospects were sometimes made to vanish by an arbitrary stroke of the pen, so that happy results and promised returns yielded to the destructive consequences of ill-concerted schemes. Such conduct was more the effect of malignant caprice, than the solicitous effects of honourable intentions. (Roughley, 1823:5)

What Roughley suggested here was that a sort of interest-based class struggle was being waged between the landed and the commercial/financial interests. This battle for the control of agrarian wealth was mediated by the attorneys, who, according to Roughley, found it in their best interests to align themselves with the merchant financiers, who were on the island, rather than the absentee proprietors, their nominal employers. This struggle can be interpreted as being waged between the metropolitan elites in the persons of the planters, versus the regional elites, personified by both the commercial financiers and the estate managers or attorneys. In these terms, Roughley's analysis reveals the tensions developing between colonial West Indian and metropolitan interests.

Roughley outlined another manifestation of this struggle when he suggested that some prudent absentees, wary of the potential abuse of power on the part of the attorneys, appointed not one but two local agents. These planters divided the responsibility of managing the plantation between what Roughley defined as "the planting attorney, the other the commissary, factor, or mercantile attorney" (Roughley, 1823:11). Roughley argued that this division of labor was not sufficient to protect the planters' interests; either the two agents were engaged in a power struggle between themselves, were in collusion to defraud the planter by falsifying the crop accounts, or else consigning the crop to a factor, who would then sell it in Kingston at a considerable profit for the agents, at the expense of the planters. In either event, the absentee planter interest was destined to lose to the machinations of the regional elites (Roughley, 1823:11–13).

Roughley attributed the cause of the planters' crisis to the tensions that existed between the regional and metropolitan elites; he logically concluded that to remedy the situation, this conflict must be mediated by what he defined as a "traveling agent." Roughley suggested that the absentees should employ agents from England to travel to Jamaica to supervise the operation of the estates. He argued that a residence of 3 months in the island per year would be sufficient time for the traveling agent to arrange the accounts for the past year, ensure that the crop be properly shipped, see to the care of the slaves and the cattle, pay the overseers and bookkeepers their proper salaries; in short, all that needed to be done by an agent could be done in these 3 months. After that term was concluded, the agent would accompany the crop back to England, ideally traveling in the same vessel in which the crop was shipped. As they would have few allegiances to the established regional elites, a new class of traveling agents would cure Jamaica of its interest-based socioeconomic crises (Roughley, 1823:29–31).

Having lived in Jamaica, Roughley recognized the existence of class-based struggles between the captive African laborers and their direct supervisors. In advising planters on the proper management of estates, he urged the overseers and plantation managers to respect the space occupied and used by the slaves, not only out of respect for them, but also for the economic interest of the estate. In choosing an overseer, for example, he entreated the planters to hire people who would not "suffer [the slaves'] provision grounds to be neglected, trespassed on, or ruined, or their houses to be out of repair or uncomfortable" (Roughley, 1823:42). Such an inconsiderate overseer, Roughley remarked, inspired the slaves to run away, which was to the benefit of no one, except possibly the escaped slave. For the well-being of the enslaved labor and thus in the best interest of the estate (from the planters' perspective), violations of the slaves' social space, i.e., their houses and provision grounds, should not be tolerated (Roughley, 1823:43–44).

Roughley illustrated the senselessness of such actions through an example with which he was familiar. According to his account, sometime around 1804 or 1805, the agent of an estate hired an overseer specifically to punish a group of runaways. The overseer lived up to his ferocious reputation by burning their houses and destroying their crops in the provision grounds. On learning of his actions, the runaways returned to the estate by night and killed the overseer. Thus, argued Roughley, the violation of social space, and the destruction of what could be considered the private property of the slaves, led not to their return to discipline, but to mortal revenge.

In the same vein, Roughley argued that overseers should not use the internal space of the elite housing to elevate themselves too much above the "white people under him," i.e., the bookkeepers and other salaried Europeans employed on the estate. According to Roughley, the overseer should:

> suffer them to sit, after business hours, in his company, instead of morosely banishing them either to their own sleeping-rooms, or to a distant dark part of the house, till mealtime is announced, which induces them to take gross freedoms with the slaves in the house. (Roughley, 1823:50)

He disapproved of the use of space to elevate and reinforce the position of the overseer at the expense of the other white employees, arguing that it created tension on the plantation, resulting in the resignation of respectable employees.

The need to employ a stable plantation staff was particularly critical for the planters. In an attempt to stabilize the European population on the plantations, a series of so-called deficiency laws required planters to maintain a minimum ratio of employed whites to enslaved Africans; those who were found in violation of this law would be assessed a fine (Green, 1976:16; Holt, 1992:125). As the plantations were required by law to employ a minimum number of whites per capita slaves, the overseers were thus forced "to accept the services of vagabonds, because few respectable young men will live with him" (Roughley, 1823:51).

While Long, Collins, and Roughley each proposed a program for the maintenance of the West Indian political economy, all three were in agreement that the region was experiencing a crisis. Taken as a whole, the discourse revealed in the treatises written by these three men suggests that members of the planter class perceived that they were facing troubled times. Significantly, each recognized that the restructuring of social space could mitigate the crisis, be it through the redistribution of land in the interior, the stabilization of population demographics through a reconstituted division of labor, or the enforcement of spatial segregation on the plantations. These tracts suggest that those experiencing the political economy of the early nineteenth century recognized the relationship between space and social relations.

CONCLUSION

Social theorists, historians of the Caribbean, and contemporary analysts agree that Jamaica was facing a crisis in the late eighteenth and early nineteenth centuries. The crisis was a result of the decline in importance of the West Indian sugar sector in the political economy of the British Empire. Facing a structural crisis, Jamaican sugar planters saw the price of their commodity fall, experienced tremendous social unrest among the enslaved labor force, and eventually were faced with a dramatic increase in the cost of labor following emancipation. This crisis led to eventual structural changes in the ways labor was exploited, and the ways that land was utilized for the production of commodities. A significant development that arose out of this crisis was the diversification of the Jamaican political economy by the introduction of new commodity production that challenged the hegemony of the sugar planters. One method by which this hegemony was challenged was the introduction of coffee production.

Having framed the context of the crisis in the political economy of Jamaica, I will now turn to frame this study in more detail by providing a contextual analysis of the sociospatial milieu experienced by the majority of people living through this crisis period: the Jamaican plantation.

4 The Focus of Analysis

Coffee Plantations in the Yallahs Drainage

I consider my district, the parish of St. David, to be in a state of tranquility. The coffee crop here, which is of considerable extent, is nearly saved: it has been got in with much satisfaction; the quantity picked per day by each person employed, when the coffee was fully ripe, being equal to that ever required.
— Patrick Dunne, stipendiary magistrate, 1835

INTRODUCTION

To examine how global historical processes were affecting change in a localized region, and how spatial manipulations were integrated into such change, the remaining chapters of this book examine the negotiation of space in the Blue Mountains of Jamaica. This chapter serves as an introduction to the area chosen as the focus of that regional analysis by outlining the structure of Jamaican plantation society and the general history of coffee production in the Yallahs River drainage, concentrating specifically on the individual plantations incorporated into this study. The Yallahs River drainage of eastern Jamaica was chosen as the project study area for several reasons. First, because the coffee plantation complex of this region was first established during the period of crisis outlined in earlier chapters, tracing the spatial transformations that occurred in this region allows me to consider how planters were adapting to the climate of crisis that affected Jamaica throughout the first half of the nineteenth century. Second, ample cartographic, documentary, and archaeological records exist to provide sufficient data for such an analysis of the Yallahs region.

SOCIAL STRUCTURE OF JAMAICAN PLANTATIONS

To best understand how plantation spaces mediated the negotiation of social relations in the nineteenth-century Yallahs region, it is important to first discuss the social hierarchy of Jamaican society. That hierarchy was defined by social divisions cut along class, race, and gender lines. Each of these categories of social relations will be discussed in turn. This outline of the general structure of Jamaican society will be followed by more detailed discussions first of the Yallahs region as a whole and then of the several plantations analyzed in subsequent chapters.

The Class Structure of Jamaica

Following the models utilized by Paynter (1985) and Wallerstein (1974, 1980, 1989), I have defined several social classes appropriate to the analysis of the Jamaican political economy: core elites, regional elites, and direct producers. Although these are heuristic categories, negotiating class identities was a very real experience for the people who lived during the study period. These terms are heuristic in as much as no one in 1820 would have defined him- or herself as a member of a "core elite" or as a "direct producer." Significant social distance did, however, exist between members of these various classes, a social distance mediated by space and power that would have clearly been recognized by people living in nineteenth-century Jamaica.

Throughout the history of the island, the great majority of manual work has been performed by people of African heritage. From the late sixteenth century, when Jamaica was first transformed into an agricultural colony by the Spanish, until 1834, when the British Parliament enforced the Emancipation Act, this class of direct producers labored under a system of slavery. During these centuries, direct producers were not free to sell their labor power in an open market, but were constrained by a regime under which humans were considered commodities valued by their potential to provide labor power to commercial agricultural estates. For the purpose of discussion, this group of people will be termed "the enslaved" or "enslaved workers," reflecting their status under slavery. These terms will be used alternately with "African Jamaican," a term that will be used more commonly to describe the direct producers of African descent after slavery was abolished in 1834.

During the period under consideration, the political economy of Jamaica was controlled by an oligarchy of European descent, primarily male and British, who owned the primary means of production. From the time Jamaica was conquered by the British in 1655 until the second half of the twentieth century, this oligarchy created wealth and power through the control of the production and distribution of sugar and other agricultural commodities. This group of regional elites is traditionally defined in the historic literature as the "planters" or the "planter class." Although some members of this social class were extremely wealthy and powerful, not every member of the British diaspora in Jamaica was affluent, nor did every planter who made a fortune in the sugar trade remain in Jamaica. Many returned to Great Britain to continue the process of wealth accumulation from there (Higman, 1984; Pares, 1950; Sheridan, 1974).

The planter class was not a homogeneous group nor was it free from internal division. For the purpose of this study I consider the planter class to have been composed not only of the powerful oligarchs, but also of those who were employed by them as estate staff. This class also included merchants, attorneys, financiers, and other members of Jamaican society involved in the commodities trade and associated service industries. While this analysis focuses on those members of this class directly involved in production, those who might be most accurately called planters, it should be noted that merchants and financiers played an extremely important role in shaping the Jamaican economy during the crisis of the nineteenth century and were occasionally at odds with the planters (Butler, 1995; Checkland, 1957; Davis, 1975; Lobdell, 1972; Pares, 1950; Roughley, 1823; Stinchcombe, 1995).

Members of the planter class most closely associated with the production of agricultural commodities are subdivided by their social ranking within their class, particularly as that ranking applied to an individual's relationship to the means of production. Both the primary and secondary literature of the historic West Indies distinguish the following members of the planter class by their relative status within the operating social hierarchy: (1) proprietors, (2) attorneys or factors, (3) overseers, and (4) bookkeepers (Roughley, 1823; Sheridan, 1974). While I am simplifying reality by considering all of these people to be members of the planter class, it should be noted that more refined class distinctions existed between some members of these groups. Such class distinctions were neither universal nor rigid, however, as it was quite possible for a young bookkeeper to become an overseer, and for an overseer to eventually become a plantation proprietor.

Proprietors, as the term itself implies, were members of the planter class who actually owned property. In Jamaica a further distinction was made between absentee proprietors, who owned estates but chose to reside in Great Britain, and resident proprietors, who either lived on the Jamaican estates that they owned or maintained a residence in Kingston or another of Jamaica's towns. Attorneys or factors were not necessarily lawyers. This latter group was responsible for managing estates for absentee proprietors; hence, they had the power of attorney over the estates that they managed. Attorneys often managed more than one estate, and thus did not necessarily live on the estates that they managed. These men were often merchants or financiers entrusted by the absentee proprietors to manage the financial affairs of their estates (Roughley, 1823).

Whether an estate was owned by an absentee or resident proprietor, the owner or manager usually hired an overseer responsible for the day-to-day operations of the estate. Each estate would have its own overseer responsible directly either to the resident proprietor or to an attorney. The overseer would be assisted in his supervisory capacity by one or more bookkeepers. Although the term implies that bookkeepers kept track of the estate's finances, which no doubt some of them did, the term is somewhat misleading. A bookkeeper was really more of an assistant overseer. A bookkeeper, usually a young man sent to an estate either from Europe or from another estate, was in many ways an apprenticed overseer. The bookkeepers would be responsible for some of the more onerous or tedious duties on the estate, like supervising night shifts and procuring provisions for the white estate staff. Just as an overseer might aspire to becoming an estate attorney or proprietor, a bookkeeper would aspire to become an overseer (Sheridan, 1974).

Taken together, the people who filled these various occupations and social categories comprised the planter class, or regional elite. Although I will use the term planter or planter elites throughout this analysis as a catch phrase for this regional elite, it should be understood that by no means was there any definition of social or material equality experienced within this class. Each member of the planter class held a position in a carefully constructed social hierarchy, mediated by differential access to power and resources.

Although the Jamaican economy was driven by agricultural production, and thus produced a primarily rural society, there were several large towns that supported what might be considered a middle class of artisans, local merchants, professionals, and others who provided support services for the plantations. This group, technically not an elite group, was not directly involved in the production of agricultural commodities, and included what Heuman defines as "free coloreds," "free blacks," and some "whites" (Heuman, 1981).

Many of the affairs of the Jamaican political economy were either dictated or mediated from Great Britain. The members of the metropolitan power structure, including the government, merchants, and absentee planters involved in the affairs of colonial production, are defined for the purposes of this study as core elites.

Racial Distinctions

African slavery created one of the most widespread and intense systems of racial segregation and oppression experienced in the history of the world. Much of one's identity in Jamaica was defined by the racial category into which one fell. In Jamaica, as elsewhere in the British colonial world, racial identity was phenotypically defined, primarily based on the relative melanin content of one's skin. The population of Jamaica has traditionally been divided into three racial categories based on this factor. Europeans and Jamaican-born children of European parents were defined as "white." Those with any phenotypic indication of mixed African and European heritage were defined as "brown" or "colored" or occasionally as "sambo" or "mulatto." Those whose phenotype was identified as purely African were defined as "black." In later decades, particularly following emancipation, a number of indentured workers from the Indian subcontinent were brought to work in Jamaica. Because these people did not fit into the existing racial hierarchy, a new category was applied: "coolie." In Jamaica, to be defined as a "coolie" was to be defined as having Indian heritage (Heuman, 1981; Higman, 1984; Stinchcombe, 1995).

Gender

Gender is a primary structuring principle by which members of a given society learn behaviors specific and appropriate to their sex. While biological sex and sexual practices are components of gender identities, gender is not dictated by biology alone. Gender is largely defined through behaviors and attitudes taught to male and female children; one must learn to be a boy or a girl. As is now well understood in anthropology, however, the correlation of biological sex and gender identity is not absolute; there are societies in which more than two gender categories exist.

It is not currently possible to suggest that more than the two conventional gender categories existed in nineteenth-century Jamaica. The historic record suggests that society was divided into males and females, although this may of course be an artifact of the androcentric bias of the male chroniclers of Jamaican history. Nevertheless, the primary documents reveal that under slavery, the planters demanded that men and women perform tasks specific to their gender. Furthermore, as women are biologically capable of bearing children, under slavery their identities were defined not only by their ability to provide labor to the estates, but also by their ability to provide laborers to the planters as well.

Creoles

One final social distinction should be briefly summarized. Those who were born in Jamaica, whether of African or European descent, were known as "creoles." In Jamaica this term was used less as a marker of race than of acculturation. It was thought that enslaved "creoles" made more docile workers, better able to withstand the strenuous demands of slave labor. Although this term was used to differentiate Jamaican-born slaves from those captured and brought directly from Africa, it was also a term members of the planter class used to define themselves, to differentiate themselves from newly arrived greenhorn Europeans, many of whom found the social and environmental climate of Jamaica extremely difficult to endure. One's identity as a creole was thus not directly tied to class, gender, or racial categories.

THE SOCIAL DIVISION OF LABOR ON JAMAICAN PLANTATIONS

Just as the planter class was divided into a number of strata based on occupation, the planters created a hierarchical social structure and imposed it on the enslaved African Jamaicans working on plantations. On each type of Jamaican plantation, whether sugar, coffee, pimento, or livestock, there were various tasks that needed to be accomplished, and a number of different kinds of skills that needed to be mastered by the enslaved workers. Roughly speaking, the division of labor on Jamaican plantations created yet another layer of social distinction, as skilled workers and some domestics gained greater access to material resources than their brethren in the fields. The task of supervising work gangs gave some among the enslaved population greater access to power on the estate, while simultaneously creating distrust and disdain for drivers who colluded with the planter class (Sheridan, 1974).

Nearly all skilled positions on Jamaican plantations went to men; often to enslaved members of the "colored" population. Plantation artisans included masons, carpenters, sawyers, and doctors. Supervisory occupations such as driver were also primarily the domain of men. Far fewer skilled occupations were open to women under the slave regime; thus, the division of labor was such that women performed more of what might be considered the menial plantations tasks — what the plantation commentator P. J. Laborie called "the works of drudgery" (Laborie, 1798). There were only two skilled occupations on most plantations open to women: nurse and midwife. Both occupations were concerned with health care-giving, and have been defined by Marietta Morrissey (1989) as "semiskilled." Other specialized tasks performed by women included many domestic tasks such as laundering for the white estate staff and cooking for the field workers (Morrissey, 1989).

By far the most common occupation for both men and women on Jamaican plantations was field laborer. Many Jamaican estates used what was called a gang system. The work force was usually segregated into three or more gangs, segregated by age and strength, and assigned different tasks. Children and adolescents were assigned to their own gangs and given tasks requiring less body strength than those assigned to the adult gangs (Higman, 1984:166). Children probably began to work in what was called the "small gang" or "children's gang" as early as age 5 or 6 (Higman, 1984:189; Roughley, 1823:108). When they reached this age and could contribute to the economy of the plantation, their status (in the eyes of the planters) was upgraded from the category of "unserviceable children" and they were moved into the working population. Higman suggests that throughout the West Indies, children aged 5–12 would work in the third gang; adolescents aged 12–18 would work in the second gang alongside weaker adults, and healthy adults aged between 18 and 45 would work in the first gang (Higman, 1984:166).

Division of Labor on Coffee Plantations

In his instructive treatise, *The Coffee Planter of Saint Domingo*, which he dedicated to the planters of Jamaica, the Haitian emigré planter P. J. Laborie provided aspiring Jamaican coffee planters with a model for a division of labor to be established among the slave population of a working coffee plantation. Not surprisingly, this division of labor was designed to emulate the emerging European class structure while simultaneously building on the precedents set by the working of the slave labor systems of the eighteenth century. The roles slaves played in the plantation economy were defined by a clearly articulated division of labor based on the type of labor expected to be performed (e.g., skilled versus unskilled labor), age, and gender. Laborie suggested that a properly working coffee plantation should have a class structure composed of artisans or skilled laborers, which he calls "artificers," as well as drivers, coffee-men, pruners, doctresses, various supervisors, and the unskilled laborers, which he calls "the gang in general." According to Laborie, the group of artificers — the skilled laborers — should include carpenters, tilers, masons, and saddlers (Laborie, 1798:164).

Capitalist class relations are, of course, hierarchically based. This was not lost on Laborie. He suggested that coffee planters should appoint slaves to the supervisory position of driver. A good driver, he argued, should be loyal to the planter, not only supervising the workers during the day, but also reporting all off-duty occupations and activities to the master. Drivers should be materially and symbolically separated from the rest of the population, and should be given a whip to carry as an ensign of their authority as intermediaries between the white planter class and the enslaved population (Laborie, 1798:165–166). Other supervisory positions on coffee plantations included mule drivers responsible for transporting coffee from the estate to the wharfs or markets from which it was to be exported or exchanged and "coffee-men" whose duty was to supervise the processing of coffee once it was brought into the mill from the field (Laborie, 1798:166). The only skilled position open to women in Laborie's division of labor was that of "Hospital Matron" or "Doctress," a woman who would supervise the care of the sick and provide obstetric care to pregnant women (Laborie, 1798:167).

The division of labor was to be assigned not only by class and gender category, but also by age. Laborie distinguished three positions to be occupied by the elderly: supervising underage children, keeping poultry, and guarding the estate's provision grounds (Laborie, 1798:168). Age categories would also be used to organize labor gangs. Laborie argued that the population should be divided into at least three gangs. One gang, made up of smaller children, would be assigned the specialized tasks of weeding and pruning (Laborie, 1798:166). Another gang, responsible for lighter work such as weeding and gathering coffee, was to be composed of young adolescents between the ages of 12 and 16 (Laborie, 1798:175). The great gang, according to Laborie's scheme, would be responsible for the bulk of the field work. Once elevated into this most directly productive labor group, adolescents would be considered adults. Laborie suggested that this status be affirmed by giving the young adults houses, provision grounds, utensils, and two hens (Laborie, 1798:175).

Having outlined Laborie's labor model, the question now turns to how much of this proposed division of labor was actualized in the Yallahs region. The best source of data to compare with Laborie's theorized class structure is an account book from Radnor coffee plantation, located in the parish of St. David. This document records the day-to-day workings of the plantation from January 1822 through February 1826.

This estate record of Radnor plantation indicates that the age- and gender-based division of labor proposed by Laborie in the late eighteenth century was actually in force in the coffee-producing parish of St. David. In January 1825, labor was organized on the gang system proposed by Laborie. The slave list appearing in the journal indicates that Radnor field workers were organized into three gangs. The first gang was composed of 85 workers, of whom 3 were drivers. Of the 85, 33 were men and 52 were women. The second gang consisted of 23 people, 14 males (including a driver) and 9 females (including a cook). The third gang was composed of 13 males (including a driver) and 9 females (including a cook; see Figure 3).

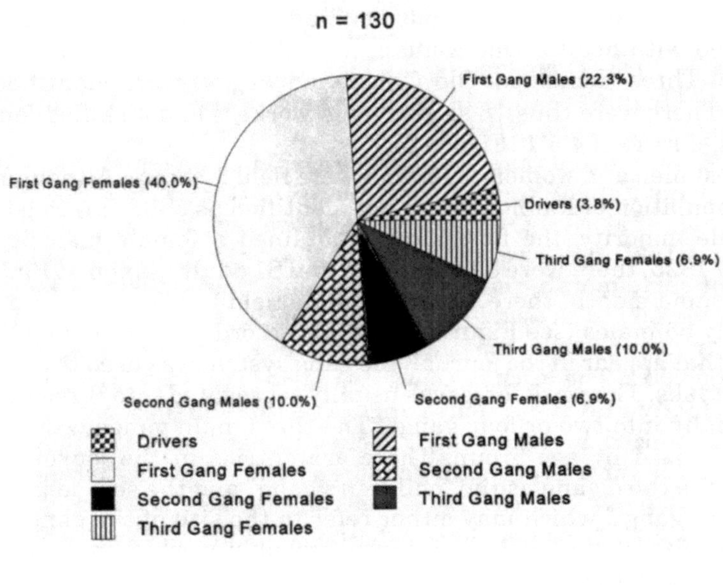

Figure 3. Division of field labor at Radnor plantation in 1824. The figures in parentheses indicate percentage of field worker population.

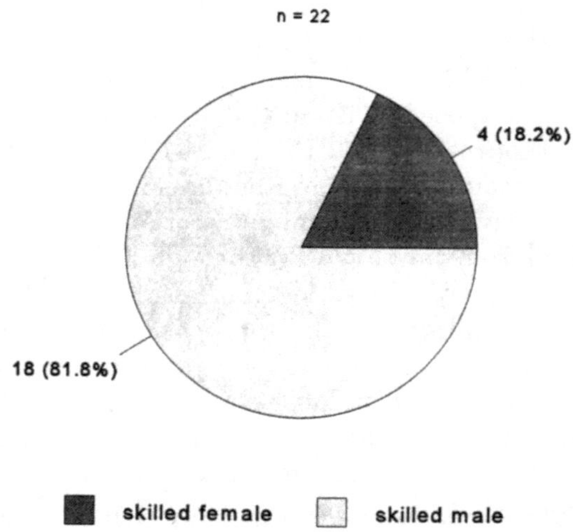

Figure 4. Gendered division of skilled labor at Radnor plantation, 1824.

The vast majority of skilled positions on Radnor were occupied by men: seven carpenters, one doctor, a boatswain, three masons, a saddler, and five sawyers. According to the plantation book, there were only two specialized occupations on Radnor open to women, both concerned with health. One woman, named Mary, was identified as a midwife. Three (Penelope, Old Chloe, and Peggy) were identified as nurses. There were thus 18 skilled male workers and 4 skilled female workers, a ratio of 4.5:1 (Figure 4).

Most men and women at Radnor were field laborers. Although the adult population of Radnor (those not identified as children) consisted of a male majority, the field gangs contained a female majority. In January 1825, there were 85 adult men and 81 adult women at Radnor. In that same month, there were 70 female field workers as compared with only 55 males (see Figures 5 and 6). According to the daily labor reports that appear in the journal, the gang system was used to allocate specific tasks. The work force was usually segregated into three gangs, occasionally into two or four gangs. The three main gangs were probably organized by age group. There are entries in the journal that identify the third gang as the "children's gang" and the second gang as the "small gang," which may either refer to the size of the gang itself or to the size of the adolescents who probably worked in that gang (Higman, 1984:166).

Although the plantation book does not provide any specific information about the ages of the work force, there is no reason to doubt that

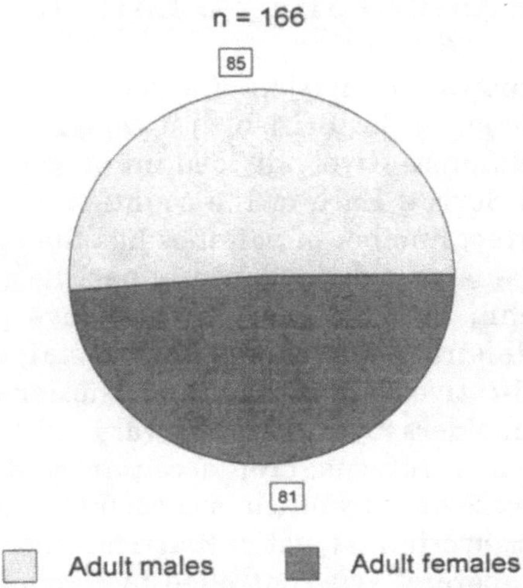

Figure 5. Adult population on Radnor plantation, 1825.

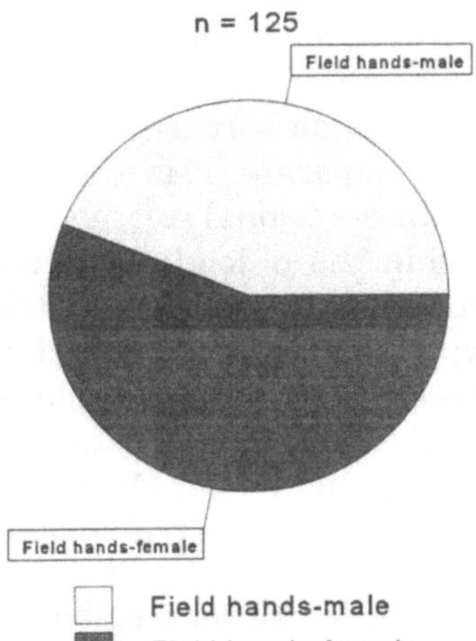

Figure 6. Gendered division of field labor at Radnor plantation, 1825.

children began to work in the third gang as early as age 5 or 6. This was a common age for children to join the productive work force in early nineteenth-century Jamaica (Higman, 1984:189; Roughley, 1823:108). All of this evidence indicates that the division of labor for coffee plantations advocated by Laborie was utilized on Radnor plantation.

DEFINING THE YALLAHS REGION

The administrative division of space in Jamaica has remained relatively stable since the British first conquered Jamaica in 1655. The island is administratively divided into three counties: Cornwall, Middlesex, and Surrey. Each of the counties is divided into several parishes. The exact number of parishes has fluctuated through time; while modern Jamaica is divided into 14 parishes, prior to 1867 there were 21 (Higman, 1988:3). The parishes have remained the most important administrative division within Jamaica. In the nineteenth century, the legislative body of the island, the Jamaica Assembly, was elected by landholders by parish (Edwards, 1810:419). Marriages, deaths, taxes, slave returns, crop accounts, and a myriad of other public records were kept by officials chosen from among the planters within each parish; criminal justice (particularly concerning enslaved and later free laborers) was controlled by magistrates appointed by the planters of each parish. Within each parish were a number of estates — private productive land holdings that, prior to 1834, were worked by enslaved Africans.

In the nineteenth century, and still today, there was another, though less formal, division of space by district. Within each parish, and sometimes crossing parish borders, the laborers and the planters shared a sense of place separate from the official division of space. There would, however, be occasional references to these more informal divisions of space within the official documents; hence one can see references to the Port Royal Mountains, a district in the contemporary parish of St. Andrew; in eastern Jamaica today, it is common to hear people refer to districts surrounding geographical or geological features (e.g., Guava Ridge) or the remnants of estates that may have become villages or towns (e.g., Epping Farm, Mt. Charles). While there is no direct historical record of the Yallahs region being defined this way, the vestry records, and the reports of later stipendiary magistrates, indicate that the coffee plantations in this region were considered as parts of two contiguous districts, the Port Royal Mountains District and the Upper Saint Davids District. The districts were divided by the parish boundary, which, for some length, was defined by the Yallahs River.

The Yallahs region includes the upland areas of two now-defunct parishes, Port Royal and St. David. Each of these parishes was abolished in 1867 during a restructuring of the island's internal political boundaries. All of Port Royal and some of St. David were incorporated into the parish of St. Andrew, while most of St. David became part of the modern parish of St. Thomas. Topographically, the study area is

Figure 7. Location of the Yallahs study area.

defined by the drainage of the upper Yallahs River, including its tributaries the Clyde, Fall, Negro, and Banana Rivers. The region is bounded on the south by the Banana River, the east by the Queensbury Ridge, the north by the Grand Ridge of the Blue Mountains (which serves as the north–south boundary between the modern parish of Portland to the north and the parishes of St. Andrew and St. Thomas to the south). The area is bounded on the west by a series of hills running from the northwest to southeast, on the western bank of the Yallahs River (Figure 7).

Historically, there have been a number of coffee plantations located within this area. In a series of reports on the condition of provision grounds published in 1836, 38 coffee plantations were reported to be operating within the Yallahs region; 26 of these were located in Upper St. David, while the remaining 12 were located in the Port Royal Mountains (PP, 1836:48, 290–293). Both parishes contained lowland and highland areas, divided by space and production. Some sugar was grown in the foothills and lowland margins, while the highlands were dedicated nearly exclusively to the production of coffee.

In terms of acreage and population, Port Royal and St. David were relatively small parishes. By extension, the Yallahs region, containing the uplands of these two parishes, was also a relatively small part of Jamaica. Through an analysis of the records of the Jamaica Assembly, Higman (1976:255–256) has calculated the size of the enslaved and apprenticed populations, by parish, for the island between the years 1800 and 1838. For most of this period the relative size of the enslaved population in Port Royal ranked eighteenth or below of 21 parishes; St. David never ranked above seventeenth for this period. In 1810, when the size of the enslaved population of Port Royal reached its peak of 7749, it accounted for a mere 2.5% of the island's total population. The population of St. David peaked somewhat later, in 1820, at which time the 8061 people accounted for 2.4% of the population. In comparison, in 1810 the enslaved population of Jamaica's most populous parish, St. Thomas-in-the-East, numbered 26,734, or about 8.5% of the island total. In 1820, Trelawny was the most populous parish with an enslaved population of 28,774, or about 8.4% of the population.

Geographically, neither parish was very large by Jamaican standards. The *Jamaica Almanac* for 1818 reported that Port Royal contained a total of 24,584 acres of land, or about 1.1% of the entire acreage calculated for the island. St. David at that time contained 46,218 acres, or about 2.1% of the island's total land. Only two parishes, Kingston and Portland, were smaller in size than Port Royal; only Kingston, Portland, Port Royal, and St. Dorothy were smaller in total acreage than St. David. In contrast, the largest Jamaican parish, St. Elizabeth, contained 217,664 acres of land, almost exactly 10% of the island's total.

Calculation of population density using these figures from the *Jamaica Almanac* for 1818 reveals that the enslaved population density for Port Royal in 1818 was 0.28 person per acre. The figure for St. David was 0.16 person per acre. The population density for the island as a whole at this time was 0.15 person per acre; thus, the population density for Port Royal was somewhat higher than the island total, while St. David did not deviate far from the average. At this time the most densely populated parish (excepting the urban parish of Kingston) was Portland, where there was a density of just under 1 person per acre of land.

Coffee production was introduced to the Yallahs region in the 1790s. Although some coffee was grown in the uplands of most Jamaican parishes, the bulk of production was carried out in mountainous regions in the parishes of Manchester and St. Elizabeth in the west and in the Blue Mountains in the east; the Yallahs region lies within the latter. Using data gathered from crop accounts filed with the island government, the Jamaican historian Kathleen Montieth has quantified the rapid development of the Jamaican coffee industry (Figure 8). These data usefully illustrate the dramatic florescence of monocrop coffee production in Jamaica in general and the Yallahs drainage specifically. According to her account, while in 1790 there was but one

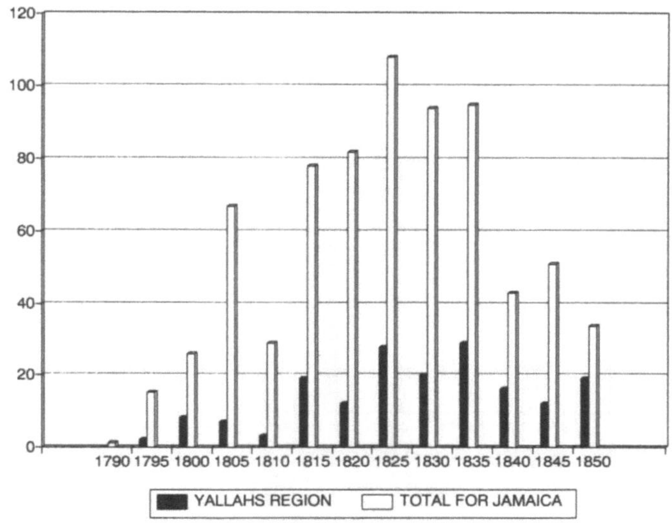

Figure 8. Productive monocrop plantations in the Yallahs region and in Jamaica, 1790–1850. Source: Montieth (1991:34–35).

monocrop coffee plantation on the entire island, by 1825 there were 110. Equally dramatic is the rapid disappearance of these plantations from the historical record after this zenith, particularly following the abolition of slave labor in the late 1830s. Be this as it may, monocrop coffee production was more persistent in the Yallahs drainage than elsewhere (see Figure 8). By 1850 more than half of the plantations still operating were located in the Yallahs drainage.

A careful analysis of the Accounts Produce, or crop accounts, indicates that, within the study period, the Yallahs region experienced two phases of coffee production. The earlier phase is best illustrated by the examples of Mavis Bank and Mt. Charles (Figure 9; for a detailed discussion of the various types of data analyzed for this study, see Delle, 1996). Mavis Bank is the first of the Yallahs plantations to appear in the crop accounts. The opening decade of the nineteenth century was the most productive for this plantation, as coffee production declined consistently thereafter, ceasing altogether by the late 1820s. Mt. Charles experienced a similar trend, also producing more coffee on average in the first decade of the century than any other (Figure 9). The second phase of production is best illustrated by Sherwood Forest and Clydesdale (Figure 10). Both of these estates are located deeper in the interior

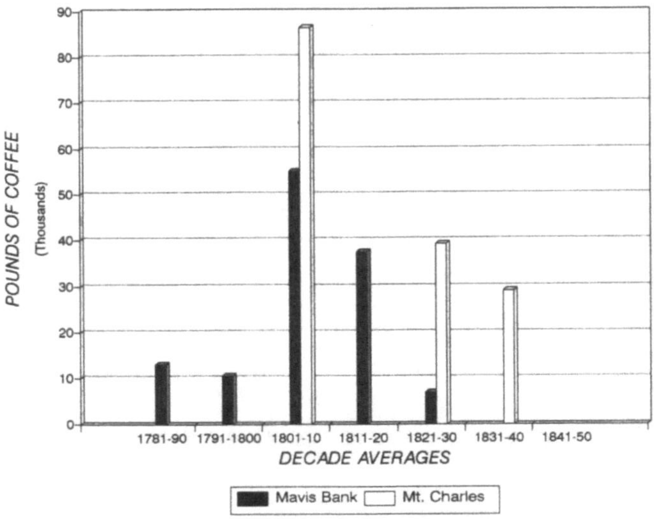

Figure 9. Coffee production at Mavis Bank and Mt. Charles, 1781–1840. Source: Accounts Produce.

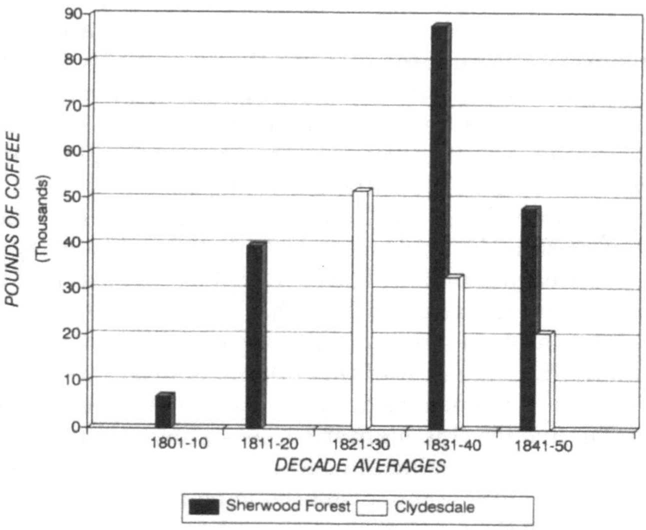

Figure 10. Coffee production at Clydesdale and Sherwood Forest, 1801–1850. Source: Accounts Produce.

than either Mavis Bank or Mt. Charles. These estates were established at a later date, and may reflect a second generation of coffee development. Given the rugged topography of the region, it is quite possible that the soils of the older estates became exhausted or suffered from erosion. As production on these estates declined, other planters may have entered into coffee production on newer estates with fresher soils. This conclusion is supported by statements made by planters concerning the nature of coffee production. In 1848, the coffee planter Alexander Geddes reported that many plantations in the postemancipation period were abandoned because they had become "worn out." According to Geddes, coffee was "planted in what is called virgin soil, and can only be once planted" (PP, 1847–1848:23/II). It is possible that these production trends were the result of poor soil management, as the planters involved in the first generation of coffee planting in Jamaica had little collective experience in coffee cultivation from which to draw advice. These early plantations may have suffered from trial-and-error experiments based on advice given from emigré planters from St. Domingue. The plantations that peaked later, including those that were still producing coffee as late as the 1840s, may have profited from the experience of the early plantations.

According to Montieth's analysis of the Jamaican coffee industry, just as the white population controlled the sugar industry, it controlled the coffee industry. Discriminatory laws passed by the Jamaica Assembly in the late eighteenth century prohibited the free colored population from owning or inheriting large estates; for example, the Inheritance Law of 1762 limited the value of property that could be transferred to black or colored people to £2000, making large property transfers to the colored children of the white planters very difficult (Montieth, 1991:70). It was not until the 1830s that the free colored population was allowed legal access to plantation ownership. Until this time, the European population had the exclusive legal right to own large export properties, like sugar or coffee plantations.

While planters shared a racial and class consciousness, the white coffee planter population was not as affluent as their counterparts in the sugar industry. Most of the people who invested in coffee could be best described as a middling class of planters, comprised primarily of merchants and estate attorneys who had been operating in Jamaica, as well as entrepreneurs newly arrived from Great Britain. Between 1780 and the early 1800s, these people, hoping to cash in during the early days of the coffee boom, borrowed heavily from Kingston merchants (Montieth, 1991:84). The difficult situation faced by these indebted coffee planters is revealed in the letters of John Mackeson, a coffee planter who arrived in Jamaica in 1807. In setting himself upon a coffee estate in the parish of Manchester, Mackeson borrowed heavily from the Kingston merchants Marsh and Fowles; by 1808 he was borrowing money from his brothers to keep himself afloat. By 1812, Mackeson wrote his brother William expressing regret in coming to Jamaica: "I wish I could get back [to England] again. I am now become fixed in this Island and must rise and fall with it" (Mackeson mss. u511/c8:3). In 1814, his debt had increased to £5000; by 1818, Mackeson had given up his life as a resident proprietor, and had returned to England (Mackeson mss. u511/c8:1–8).

Although no similarly specific information has yet been discovered for the first generation of coffee planters in the Yallahs region, many of them were probably similarly situated. Most owned only one or two properties and, with a few notable exceptions like Robert Morgan, the absentee owner of Radnor, most lived either on their estates or in Kingston.

ENCAPSULATED HISTORIES OF THE INDIVIDUAL PLANTATIONS UNDER CONSIDERATION

As the purpose of this study is to analyze the relations between historical process and spatial dynamics, and not to produce a detailed culture history of the Yallahs region, the archaeological phase of the study focused on a sample of plantations. Given the limits of the cartographic, documentary, and archaeological data, the scope of analysis has by default been limited to those plantations represented in all three of these various data sets. While presenting a detailed social history of every plantation located in the Yallahs region is not the point of this study, a brief outline of the histories of some of these estates helps to establish the context for the spatial phenomena to be considered (see Figure 11; for an extended discussion see Delle, 1996).

There were specific reasons for examining each of the following, beyond the logistical fact that archaeological, cartographic, and documentary records exist for each. Geographic coverage of the region was one consideration. Plantations from both Port Royal (Mavis Bank, Mt. Charles, Chesterfield, Clydesdale) and St. David (Whitfield Hall, Sherwood Forest) were included. The effort was made to include estates from throughout the region: Mavis Bank, Mt. Charles, and Chesterfield were located in the geographic center of the Yallahs region, near the confluence of the Green and Yallahs Rivers; Clydesdale was located in the far northwest of the study area, on the Clyde River, which is a

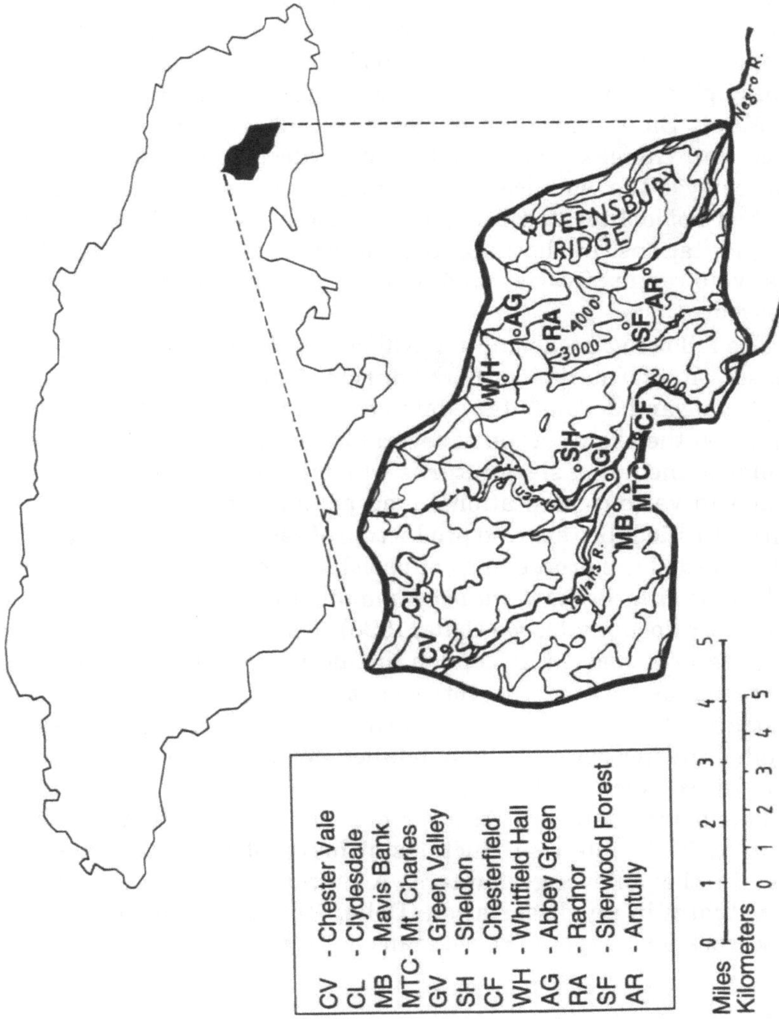

Figure 11. Locations of plantations mentioned in text.

tributary of the Yallahs; Whitfield Hall was located in the far northeastern section of the Yallahs region, at the base of Blue Mountain peak, the practical upper elevation limit of coffee production; Sherwood Forest was located in the far southeast of the study area on the Negro River, a tributary of the Yallahs. Several other factors were considered in choosing which estates to examine, mostly involving historical contingencies. For example, Sherwood Forest annexed several surrounding plantations, including Arntully, Eccleston, and Brook Lodge. Both Mavis Bank and Mt. Charles were subdivided and eventually evolved into villages. Chesterfield and Clydesdale eventually became the property of the Jamaican government; as a result, private development has not obliterated the archaeological record of these historic coffee plantations. In choosing this non-random sample, it was hoped that a wide range of variation of historic spatial phenomena could be addressed.

The following encapsulated histories of these various plantations are based on information gathered during the process of analyzing the cartographic and documentary data accrued for this study. This analysis involved the examination of several types of public record archived in Jamaica, including slave returns which listed the number of slaves attached to various plantations, crop accounts or Accounts Produce which quantified the reported production of each estate, vestry minutes which recorded the proceedings of parish governments, and local periodicals, particularly the *Jamaica Almanac* (for a more detailed discussion of these documents, see Delle, 1996). The analysis also considered colonial records housed in London, including reports made to and by Parliament, as well as the correspondence of several planters and magistrates. As a caveat, time constraints in performing this research precluded the incorporation of a detailed deed analysis in this project. In the future, such research might shed further light on the spatial phenomena analyzed here. As a further note, it was difficult to impossible to make legible reproductions of many of the estate maps discussed in the following chapters. A number of these appear in a digitized form in my dissertation (Delle, 1996); the reader is referred to that work for images of those maps not reproduced here.

Mavis Bank

The earliest documented reference to Mavis Bank, a plantation located in the parish of Port Royal, appears on a diagram of a division of patented land performed in 1773, by A. E. — either Alexander or Archibald Edgar, both of whom were surveying land in late eighteenth-century Jamaica (Higman, 1988:294). According to this diagram (STT 1453), before this date Mavis Bank was part of a larger parcel of land patented by Jacob Stoakes. In 1773, this land was divided between John Patterson, who received 67 acres (which would later become part of Mt. Charles plantation), and John Nixon, Esq., who was granted the remaining parcel of land, which the diagram represents as having 133 acres. It is Nixon's parcel that would eventually become Mavis Bank. On a later map, drawn in 1808, Mavis Bank is represented as having grown to 302 acres (Higman, 1988:167).

According to the Accounts Produce (AP 15:205), by 1789, the plantation was owned by an absentee proprietor, "John McCourtie of Great Britain." At this time, Robert Spaulding was identified as McCourtie's attorney, indicating that he was the manager of the estate in McCourtie's absence. According to a map drawn in 1808, Mavis Bank was still the property of "John Marcoutie" (STA 35). However, the Accounts Produce identify "Alexr. McCourtie" as the proprietor in 1795, and Raymond Hall as the proprietor in 1797. According to the *Jamaica Almanac*, by 1818 the estate was the property of John Biggar who held the estate at least until 1823 (*Jamaica Almanac* 1818–1823). In 1841 and 1842, the property was subdivided into a number of small lots, with two larger lots respectively purchased by Robert Sylvester (75 acres) and Dr. William Thompson (60 acres); Thompson renamed his portion of the estate "Forest Park" (STA 60). In the early 1850s, the Sylvester piece was further subdivided. By 1841, no coffee works were located on either of the two diagrams outlining the original subdivision, suggesting that the property had ceased to function as a coffee plantation by this date (STA 35, STA 60, Higman, 1988:168).

Several Mavis Bank overseers can be identified. In 1795, the same year that "Alexr. McCourtie" was proprietor, James Rathray was the overseer of the estate. In 1796 and 1797, the overseer was John Nicolson; in 1801, the position fell to John Bywater. The longest serving overseer was William Breakenridge, who filled the position from 1802 through 1815. In 1816, the overseer was *Thomas* Breakenridge, perhaps suggesting that the position was transferred within a family. The next year Robert Gibson served as overseer. George Osborne, the last Mavis Bank overseer identified in the Accounts Produce, served between 1818 and 1823.

The first reference to coffee production at Mavis Bank appears in the Accounts Produce for 1788. In that year the estate reportedly produced 13,113 pounds of coffee. The last recorded crop for the plantation was in April 1821, when only 7012 pounds were consigned to James Walker and Co. of London. In this same entry it is indicated that the overseer George Osborne had turned to "jobbing" the enslaved workers attached to the estate, that is, renting out the labor power of the workers owned by the plantation to other planters (AP 57:55). It is possible that after this date no large-scale coffee production was carried on at the estate. In 1838, the *Jamaica Almanac* published the number of apprentices attached to the estates at the time of abolition. There is no listing for Mavis Bank, reinforcing the conjecture that sometime prior to this date Mavis Bank ceased its operations.

Beginning in 1818, Jamaican planters were required to register the number of enslaved people living on each estate, as well as the births and deaths that had taken place over the course of the year. These returns were occasionally published in the *Jamaica Almanac*. In 1818, the *Jamaica Almanac* reported 106 enslaved workers residing on Mavis Bank, the property of John Biggar. In 1820 and again in 1821, the number registered was 71; it is likely that somewhere near this number of people was hired out during the jobbing phase of the plantation. Mavis Bank was subdivided into lots and conveyed to small settlers in 1841 and 1842. Subsequent to this subdivision, the estate became a village, and today is one of the chief upland towns of the parish of St. Andrew.

Mt. Charles

As is the case with Mavis Bank, Mt. Charles has been included in this study because the plantation was eventually subdivided and has become a village of the same name. Although the cartographic data do not suggest a date for the subdivision of the plantation, it is likely that it occurred in the years following emancipation, probably in the 1840s or 1850s. In the Jamaica Archives there are listings for mortgage conveyances of the plantation dated March 22, 1821, and April 4, 1838; it is likely that the subdivision occurred after this latter conveyance.

Without the use of deed research, the history of ownership of Mt. Charles is more difficult to trace than that for Mavis Bank. The earliest cartographic reference for Mt. Charles occurs on an undated diagram of land division between the Barbecue, Yallahs, and Fall Rivers. Although this anonymous map is undated, it does list several dates on which properties surrounding Mt. Charles were surveyed, but, unfortunately, no date is given for the original survey of Mt. Charles itself. The latest date appearing on the map is 1773; it is likely that this diagram was drawn sometime in the mid- to later 1770s. On this map, Mt. Charles is listed as containing 302 acres, the property of Robert Watts. Interestingly, none of the surrounding properties are given place names, but are only associated with the names of proprietors, suggesting, perhaps, that coffee production had not yet begun. If indeed this space had not yet been put into production, these lots would not have been properly considered plantations. However, within the Watts parcel, a small square is labeled "Mt. Charles"; it is unclear whether this symbol represents a working plantation or merely a house.

The *Jamaica Almanac* information indicates that by 1818 the estate was owned by Robert William Tait, who was apparently an absentee proprietor residing in London; William Harris is identified as the estate attorney for that same year. The Accounts Produce indicate that as early as 1808 Tait had been living in London; the entire crop for that year was shipped to him there. The Accounts Produce for 1820 associate the name Keyran Tobin with the estate; he was most likely the overseer. In the following year, William Harris is identified as overseer. In 1823, the estate is listed as the property of the heirs of Robert Tait; his will was listed in probate in 1822.

The first documented coffee production at Mt. Charles appears in the Accounts Produce for 1804. In that year, Mt. Charles produced 102,729 pounds of coffee (AP 33:59). The estate was extremely productive for several years, producing a crop of 63,756 pounds in 1805 (AP 34:61) and 141,620 pounds of coffee in 1806 (AP 36:95). From this high yield, production dropped off sharply to a mere 42,279 pounds in 1807 (AP 37:98), although it rebounded to 82,402 pounds in 1808 (AP 40:15). For some as yet unknown reason, the estate does not again appear in the Accounts Produce until 1822. By this time, the economy of the estate was diversified to include jobbing. Although the estate produced 35,691 pounds of coffee that year, the Accounts Produce list money received for jobbing at Carrick Hill estate, and hiring out the plantation's skilled carpenters to Bull Park Pen. Save for a brief rebound in 1838, the production of the estate dropped off every year following the implementation of apprenticeship in 1834, suggesting that laborers began to spend more time cultivating their own crops than in working for the estate. Production dropped to a low of 16,124 in 1839, which is the last time the estate appears in the Accounts Produce. It is likely that soon after this date, Mt. Charles was subdivided and sold to the small settlers identified in an undated map (STA 56).

Beginning in 1829, the revenue of the estate was supplemented by the renting of the plantation great house. In that year, the house was rented to Dr. G. W. Schweter. In 1830, the house was rented to a Dr. A. Dallas, who, according the Accounts Produce, rented the house from January of that year through July 17, 1832. There is no indication in the Accounts Produce whether the house was rented to another tenant after this time, or whether someone attached to the estate lived there, or whether it sat vacant.

The Accounts Produce for 1804 identify John Bywater as the overseer at Mt. Charles. He remained in this position until at least 1808, the final year Mt. Charles is listed in the accounts until 1822, when James C. Pownell, Esq. appears in the record as the probable estate attorney. In 1824, George Davis is identified as the overseer. By 1825, management of the estate fell to Alexander Reid Scott, who is listed in the Accounts Produce as the "agent of the late James C. Pownell, dec'd" (AP 63:50). By 1828, the estate had come under the control of Andrew Carey. The Accounts Produce for 1829 state the estate and "premises of Mt. Charles" were under the care and direction of Carey "of which he is in possession of as overseer." By 1831, Alexander Bizzett was in control of the estate. From 1834 through the end of the record in 1839, Mt. Charles was under the direction of the overseer John Anderson, while the *Jamaica Almanac* for 1838 lists George Wright as the attorney for Mt. Charles.

The slave returns published in the *Jamaica Almanac* for 1818 indicate that by this time there were 118 enslaved workers working and living at Mt. Charles. This number increased to 127 by 1820 and 139 by 1821. At the time of full emancipation in 1838, the number of apprentices attached to the estate was 100.

Chesterfield

Chesterfield was located on the border between the parishes of Port Royal and St. David; the industrial works as well as the households of the white estate staff were located in Port Royal while the coffee fields and the village were located in St. David. Chesterfield first appears in the historical record through an entry in the Accounts Produce for 1794. In that year, the plantation belonged to the estate of Archibald Thomson, who apparently was recently deceased. A Mr. Douglas Thomson is listed in the Accounts Produce as having sold some of the coffee. It would appear that he was temporarily managing the estate while Archibald Thomson was still alive; he is listed as "having acted for said." After Archibald's death, the management of the estate fell to his widow, Mrs. Jane Thomson; she is listed as the proprietor of the plantation in the Accounts Produce for 1795; however, her legal title to the plantation is ambiguous. The listing for 1797 reads that the estate is "under the management and direction of Mrs. Jane Thomson." She clearly was running the affairs of the plantation. If, however, Douglas Thomson was the son of Archibald and Jane Thomson, the estate may legally have belonged to him. The estate may have been in some state of confusion. Chesterfield does appear in the Accounts Produce for 1798, but does not appear again until 1810 and then again not until 1822. In 1838, the *Jamaica Almanac* listed Alexander Bizzet as proprietor of the estate. No other proprietor is listed in the Accounts Produce. However, the *Jamaica Almanac* for 1818 lists James Muir as the proprietor. The only other known proprietors are listed on a map of the plantation dated 1854. At this time, the estate belonged to Charles and George Barclay.

The record of coffee production from the Accounts Produce for Chesterfield is fragmentary. The first record, from 1794, indicates that the plantation produced 13,242 pounds of coffee. In 1795, Chesterfield produced 27,147 pounds. By 1797, production slipped back to 13,891. In 1798, production increased slightly to 21,444 pounds. There is no other record until 1810, when the plantation reported a crop of 68,910 pounds. After 1810, there is a 12-year gap in the plantation record, ending in 1822. In that year, Chesterfield reported a crop of 35,134 pounds. In 1823, production dipped again to 11,237 pounds. Production increased over the next 2 years, reaching 50,571 pounds in 1825. Chesterfield remained a productive concern during the 1820s. In 1829, the estate reported a crop of 145,831 pounds; this is the largest crop recorded in the Accounts Produce for Chesterfield. The last reported crop was in 1830, when Chesterfield produced 47,584 pounds of coffee. Whether the estate remained in production after this date is as yet unclear. However, a map of the estate dated 1854 suggests that the estate was still divided into coffee fields. Whether the estate remained in production through 1854 or returned to production around this time is as yet undetermined; however, the apprenticeship lists appearing in the 1838 *Jamaica Almanac* indicate that 55 people were still attached to the estate at this time.

By the 1820s, the enslaved population of Chesterfield was relatively large. Because part of the plantation was located in St. David, the estate appears in the Vestry Minutes for that parish. In 1801, the estate had 70 enslaved laborers working on it. In 1818, the *Jamaica Almanac* reported that 148 people were attached to the plantation. The population for 1820 is ambiguous, but probably numbered 166, climbing to 173 in 1821.

Clydesdale

The first known historical reference to Clydesdale, an estate located in Port Royal, appears on a diagram of the plantation and surrounding parcels of land drawn in 1801. Prior to this date, the parcel of land that became Clydesdale was part of another estate, Chestervale. In 1800, the proprietor of Chestervale, Dr. Colin McClarty, divided his estate into four parcels. The largest, containing 874 acres, was retained by Colin McClarty. Two sections were conveyed by McClarty to William Griffiths; these two parcels were identified as Newport Hill plantation. The fourth and final portion of the subdivision was "conveyed by Doctor Colin McClarty and uxor to Dr. Alexander McClarty," who may have been the son of Colin McClarty (STA 247). The portion of the estate that was conveyed to Alexander McClarty contained 329 acres; a diagram suggests that the industrial works of Chestervale were included in this conveyance. If this indeed was the case, Colin McClarty may have given up the idea of producing coffee in the early nineteenth century, turning the productive end of his estate over to Alexander. Chestervale does not appear in the Accounts Produce until 1821, which would have left sufficient time for the estate to have been sold and improved by another owner.

Alexander McClarty appears to have remained the proprietor of Clydesdale until his death in 1821. By 1823, the old section of Chestervale that had become known as Newport Hill plantation was attached to Clydesdale. In 1823, the *Jamaica Almanac* lists the "estate of Alexander McClarty" as owning the plantation; this same listing appears in 1826. The cartographic and Accounts Produce data reveal no further owner of the plantation; the apprenticeship statistics printed in the 1838 *Jamaica Almanac* indicate by this time the estate had become the property of Alexander Campbell.

In 1822, sometime after McClarty's death, the management of the estate fell to Alexander Campbell. By 1825, the next time Clydesdale appears in the Accounts Produce, George Dixon had assumed management of the estate, remaining in this position until 1839, when the affairs of the plantation were taken over by Robert Hogg. By 1841, Charles Lascelles had assumed management of the estate; his is the last name associated with the estate. The Accounts Produce entry for 1842 is the last appearance of Clydesdale in the record.

Clydesdale was a moderately productive coffee plantation. The largest recorded crop was that for 1829, when the plantation reported producing 77,654 pounds of coffee. Between 1821 and 1843, the Accounts Produce indicate that production fell below 20,000 pounds only three times; however, no record exists for the years 1823, 1824, 1827, and 1840. These records indicate that the management of Clydesdale was able to organize sufficient labor to successfully harvest the crop at least for the first half decade following emancipation. The slave returns for 1818 published in the *Jamaica Almanac* report 109 enslaved laborers attached to the plantation at that time. This number increased to 115 by 1820, decreasing by one to 114 in 1821. At the end of apprenticeship, there were 89 apprentices working on Clydesdale.

Whitfield Hall

The cartographic record for Whitfield Hall, which was located in the parish of St. David, is not as rich as that of the plantations in Port Royal. Only one map survives from this plantation, and not very much information is contained therein. However, the overseers house from Whitfield Hall survives pretty well intact. For this reason, and because Whitfield Hall is among the plantations located at the highest elevation in Jamaica, this estate has been included in this analysis.

Whitfield Hall appears in the Vestry Minutes for St. David as early as 1801. In September of that year, the Vestry reported that the estate was the property of Thomas Leigh, Esq., and that Leigh lived with two of his family members, his wife Ann Leigh and Sarah Leigh, probably his daughter. By March 1803, Leigh's family included Ann Leigh junior. By January 1804, the Leigh household included Thomas Whitfield Leigh and Harriet Leigh. In 1818, the elder Thomas Leigh died. As seems to have been typical of nineteenth-century Jamaica, there was no quick transfer of the property following Leigh's death. In 1830, the Accounts Produce list the "Estate of Thomas Leigh, deceased" as the proprietor of Whitfield Hall. According to the Accounts Produce, by 1829 the estate went into receivership, with the plantation manager Francis Clarke identified as the receiver; in 1832, Alexander R. Scott and John J. Stamp were listed as receivers of the estate in the slave returns (SR 1832/128/164).

Whitfield Hall first appears in the Accounts Produce in 1817, when 5301 pounds were listed as having been consigned to the merchants John Atkins and Son of London. The Accounts Produce information for Whitfield Hall is fairly complete; between 1817 and 1847, the first and last years the estate appears in the records, only the reports for 1820, 1821, 1823, and 1824 are missing. The largest crop produced was 56,673 pounds, reported in 1827. The plantation reported a crop larger than 40,000 pounds only two other times, once in 1818 (51,449 pounds) and again in 1822 (50,065 pounds). Production fell off following the end of slavery. Between 1835 and the end of the record, the plantation produced crops of over 20,000 pounds only twice, in 1840 and again in 1847. There is no indication that coffee was produced for export after 1847.

Following the end of the apprenticeship period, the manager of Whitfield Hall, Robert J. Hall, began renting out pieces of the estate to the emancipated laborers who chose to remain on the estate. According to the Accounts Produce, in 1841 Hall collected £10.18.0 in "rents received from laborers." In 1842, the amount more than tripled to £36.0.0. It is possible that levying this rent on the laborers encouraged some of them to move on, as the amount collected decreased steadily from £25.4.0 in 1843 to £18.6.6 in 1844, £18.3.0 in 1845, and a mere £8.2.0 in 1846. The record for the following year does not indicate that any rents were collected; incidentally, this was the last year that Whitfield Hall coffee production appears in the Accounts Produce.

Through the Vestry Minutes and Accounts Produce, it is possible to trace the history of the white estate staff of Whitfield Hall. The Vestry Minutes list the number of "whites" on each of the St. David plantations. In 1801, the Leigh family was accompanied by George Bull and John Barclay on Whitfield Hall; Bull and Barclay were most likely employed as overseer and bookkeeper on the estate. In 1804, Gabriel Gummer is the only other white in residence. The next appearance of an overseer's identity is the Accounts Produce entry for 1817, when Francis Clark signed the register for Whitfield Hall; in 1819, Clark is identified as "planter," a title commonly used by both overseers and estate managers during this period. By 1825, James Russell is identified as the overseer for the plantation. Clark's name also appears in this entry; the coffee produced in that year was reportedly "delivered to Francis Clark, Esq." Because Clark is identified as "Esq.," it is likely that he was by this time the attorney or manager of the estate. In 1826, Clark again appears as the only name associated with the estate. In 1827 and 1828, James Lernancy was employed as the overseer for the estate. The following year, James Robert Ramsey served as the overseer; Clark was identified as the receiver of the estate. Ramsey served until 1832 when he was replaced by Alexander R. Scott. Joseph Ornsby replaced Scott in 1833. In 1835, he was replaced by James Williamson Crosbie, who in turn was replaced by Charles Mapother. Mapother was employed until 1838, when the affairs of the estate were taken over by Robert J. Hall. Hall served in 1838 and then again from 1840 through 1844; A. Murray managed the estate in 1839. John H. Hall, Benjamin Magill, and J. Davidson served in 1845, 1846, and 1847, respectively.

According to the St. David Vestry Minutes, in December 1801 the enslaved population of Whitfield Hall stood at 75. By the end of 1805, that number had officially increased to 87. By 1818, the population had increased to 103, decreasing to 98 in 1820 and rebounding to 100 in 1821. At the end of apprenticeship, 58 laborers were attached to the estate.

Sherwood Forest

As was the case with Whitfield Hall, the cartographic record for Sherwood Forest is not very extensive. Sherwood Forest does, however, appear in the Accounts Produce as early as 1801. In that year, the Vestry Minutes identify William Thompson as the only white occupant of the plantation. The first owner who can be confidently identified is Thomas Cuming, who is listed as such in the *Jamaica Almanac* for 1818; he is, however, also identified as having been deceased at that time. The Accounts Produce for 1801 and 1802 associate a Robert Ross of Kingston with the plantation; Ross may have been either the proprietor or an attorney for an absentee proprietor. A surveyor's diagram dated 1809 indicates that at this time the plantation contained 367 acres (STT 1439).

In 1821, the estate became the property of William Rae, who was the owner of the abutting Arntully plantation. Rae was the proprietor of Sherwood until his death on May 6, 1837. Following Rae's death, Sherwood was devised to his heirs who were responsible for the plantation at least through 1852 when the Accounts Produce indicate that the crop for that year was shipped by the orders of the "devisees" of Rae's estate. The only devisee identified by the Accounts Produce is Adam Newall, Esq., who is mentioned in the entry for 1851. In that same year, John Hall and Company were identified as "agents for the devisees of the estate of William Rae." It seems likely that an employee of this firm was actually managing the affairs of the estate.

There are significant gaps in the Accounts Produce record for Sherwood Forest. No data exist for the years 1803–1814, 1818–1836, and 1841–1846. The first reported crops from Sherwood Forest were modest. In 1801 and 1802, the plantation reported crops of 5679 and 7758 pounds, respectively. From 1815 through 1852, the plantation reported no crops smaller than 30,000 pounds; however, there are 23 years missing in the record during this span; it is possible that there were smaller crops that went unreported. The largest crop was reported in 1838 — 108,293 pounds. Sherwood Forest seems to have prospered during the apprenticeship and immediately following emancipation; in 1837 and 1840, crops of 82,644 and 72,294 pounds were reported. These large figures may represent a prosperity brought on by land speculation; the Accounts Produce suggest that sometime prior to 1838, the trustees of Sherwood Forest had acquired two abutting plantations, Brook Lodge and Eccleston. Unfortunately, the crop data for Sherwood Forest for the years immediately prior to emancipation do not exist, thus we cannot determine whether these large crops represented an increase or decline in production from this earlier period.

Although the Accounts Produce for both 1801 and 1802 were reported by Robert Ross, he does not appear to have lived at the plantation. The Vestry Minutes report William Thompson as the sole white inhabitant of the plantation in June 1801. In December 1802, James Drysdale was Sherwood Forest's only white inhabitant; in 1803 and 1804, Thomas Firby was the only white inhabitant. It is likely that all of these men served as overseers under the direction (at least in 1801 and 1802) of Robert Ross. The entry in the Accounts Produce for 1815 — the first following the first gap in the records — identifies Morto O'Sullivan as the overseer for the estate. O'Sullivan remained in this position at least through 1817, the entry immediately preceding the second gap in the records. Following this gap, the record for 1837 identifies Andrew Murray as overseer. Murray remained as overseer at least until 1841, the last date preceding the 1841–1846 gap in the records. By 1838, Murray was in charge not only of Sherwood Forest, but also of Brook Lodge and Eccleston plantations, which are described in the Accounts Produce for 1838 as "appendages" of Sherwood Forest (AP 82:79). When the record resumes in 1847, John Fogarty is listed as the overseer of "Sherwood Forest, Brook Lodge, Eccleston and Arntully plantations" (AP 91:211); Sherwood Forest seems to have continued to acquire land in the years following emancipation, perhaps explaining the bumper crops reported for the late 1840s and early 1850s. In 1848, Edward Mais was identified as the attorney of the property while Hugh Fraser Leslie was the purported proprietor — it is unclear whether or not Fogarty was still employed at this time. Although no one besides Mais is identified in the remaining entries (through 1854), it seems likely that Fogarty or another overseer was employed at Sherwood Forest, as Mais was by that time residing in Kingston (AP 95:106; PP, 1848:23/3, 167).

The size of the enslaved population during Sherwood Forest's early years reflects the modest production reported for this time. The Vestry Minutes for 1801 report an enslaved population of only 24; by 1818, this number had increased to 82. Following the death of Thomas Cuming (date as yet unknown), Sherwood Forest seems to have experienced a period of rapid expansion. According to the slave returns for 1820 and 1821 published in the *Jamaica Almanac*, the population jumped to 124 and 139 respectively; the only conceivable ways the population could have increased so rapidly are through the purchase of enslaved laborers from other Jamaica planters or through the purchase of illegally smuggled Africans, as the slave trade had been abolished by Britain in 1807.

William Rae, the proprietor of the estate following the death of Cuming, was, by the time of his death in 1837, in possession not only of Sherwood Forest, but also of Eccleston, Brook Lodge, River Head, and Arntully plantations. At the end of apprenticeship, Sherwood Forest alone had 121 laborers attached to it; the other four estates had a total of 370 apprentices working on them (*Jamaica Almanac*, 1818:26).

The analysis that follows in Chapters 6 and 7 focuses primarily on this sample of Yallahs estates. However, these represent only a fraction of the dozens of coffee plantations located in the region in the nineteenth century. As the situation warrants, several other estates that will be mentioned in the text, notably Arntully, Sheldon, Abbey Green, Ayton, New Battle, and Green Valley (see Delle, 1996, for a more complete list of Yallahs estates).

CONCLUSION

The quantity and quality of the cartographic, material and archaeological record of the Yallahs region provide an excellent data universe with which to analyze the transformations in space that were occurring during the crisis of the late eighteenth and early nineteenth centuries. As far as can be determined from the historical record, each of the estates in this region was established between the late 1780s and the late 1810s, the very period during which the Jamaican crisis was creating opportunities for some entrepreneurs to break the hegemony of the sugar sector. The records of these estates provide sufficient evidence to consider how productive spaces were first created, how these spaces developed into working plantations, and how these enterprises were spatially redefined once the system of enslaved labor was replaced by wage labor. In the following three chapters, I discuss not only the material spaces of Yallahs region coffee plantations, but also analyze how specific plantation spaces mediated the negotiation of those social identities discussed in this chapter, primarily racial and class identities, both prior to and following the abolition of slavery. Chapter 5 considers cognitive space through an analysis of the imagined space of eighteenth- and early nineteenth-century plantation theorists. Chapter 6 considers the social space and spatialities of coffee plantations under the regime of slavery, from the first introduction of coffee into the Yallahs landscape through emancipation, while Chapter 7 considers how these plantations, once established, were spatially transformed following the abolition of slavery.

5

Analyzing Cognitive Space

The Imagined Spaces of the Plantation Theorists

A few planters suffer their negroes to make their own huts themselves, and in what form they please; but these will always be very incorrect, and perhaps insufficient. Besides, it seems that this building of houses is one of the obligations of the master.
— P. J. Laborie, 1798

INTRODUCTION

This chapter continues my exploration of the spatial dynamics of the transition from mercantile to competitive capitalist modes of production in the early nineteenth century by exploring how a new sociospatial order was cognitively conceived by various plantation theorists in response to the crisis of the early nineteenth century. Prior to emancipation, such schemes sought to re-create a social order in which the planter class maintained exclusive control of productive space; following emancipation, the schemes attempted to materially create and legitimate a new set of social relations of production based on wage labor. In Jamaica, this spatial reorganization included the development of new patterns of landownership as the emancipated slave labor force transformed into an agrarian peasantry, the creation of new town and village forms inhabited by the peasantry, and the abandonment of large areas of land formally cultivated in sugar. Significantly, in the years leading up to emancipation, these processes also included the introduction of new crops — most notably coffee — and the spaces needed to create commodities from these plants. To understand how plantations were designed and imagined, that is, how some elements of cognitive space were negotiated, this chapter will analyze how contemporary plantation theorists recognized plantation spaces during the period of crisis I outlined in earlier chapters.

IMAGINING PLANTATIONS: THE THEORISTS, 1790–1834

Under mercantile capitalism, the exploitation of slave labor had provided Europeans with the opportunity to accumulate vast amounts of wealth. With the abolition first of the slave trade (1807) and then of slavery (1834), new forms of social inequality emerged in Jamaica. Access to economic resources, social prestige, and productive space was defined by social categories based on phenotype (race), sex/gender, and social position within the emerging class structure. The existence of race/gender/class stratification was not unique to this particular phase of capitalism; however, as social relations were being redefined during the transition within capitalism, plantation theorists reflected on how these social constructs should be manifested in Jamaica.

In 1774, just at the beginning of the period of structural crisis discussed in Chapter 3, the planter-historian Edward Long proposed a scheme by which a new social structure could be imposed on Jamaica. Long believed that this social structure could be constructed by reorganizing the existing spatial structure of Jamaica. As discussed in Chapter 3, Long believed that Jamaican society was facing impending instability because the island's racial composition was too bipolar, i.e., the ratio of whites to blacks was too low for his liking. Long argued that the social situation in Barbados, where the ratio of whites to enslaved Africans was much higher than in Jamaica, was more stable than that in Jamaica. In effect, Long hoped to replicate the Barbadian situation in Jamaica. He imagined that effective social stability could be constructed by redefining the spatiality of Jamaica, notably through the construction of small settlements or plantations, which he called townships, in interior districts where land could be purchased cheaply. A central element of this scheme was the introduction of new types of commodity production to Jamaica, using these newly defined interior townships as the locus for such production. He hoped that the creation of such settlements would encourage the development of a middle class; the existence of such a middle class would provide profitable opportunities, and thus attract small-scale investors. He believed that with a relatively small capital outlay, of approximately £1000, a small planter could initiate indigo or coffee production, notably with a work force of no more than 20 enslaved Africans. In Long's mind the creation of these settlements would result in several developments that would stem the social crisis he perceived: The wilderness of the Jamaican interior, which had for decades been the refuge of escaped slaves, would be brought under production; the vast social inequalities among the European settlers established by the hegemony of the sugar interest would be mitigated; the economy of Jamaica would be diversified; and the social structure (from his racist point of view) would be stabilized, as the ratio of whites to blacks would be increased (Long, 1970 [1774]:407ff).

Thus, Long suggested that a new class of small landholders be created as a sort of pioneer vanguard. It would be up to this middle class to organize the difficult and dangerous task of clearing the interior forests and establishing new productive lands, farther from the sea than the established sugar plantations. Presaging the ideology of free enterprise, Long suggested that if these small settlers worked hard enough, they could eventually buy out their neighbors, increase their holdings, and elevate themselves to a higher socioeconomic stratum. Those unsuccessful in establishing themselves, along with a new wave of European adventurers, would build farther inland, eventually establishing commodity production throughout the island (Long, 1970 [1774]:407ff).

Long argued that the economy of the island would benefit from the introduction of small-scale commodity production, both for the internal consumption of Jamaica and for export back to England. He argued, for example, that the encouragement of animal breeding ranches, known as pens, could save the larger planters the expense of importing livestock for food and labor, and provide smaller settlers with the opportunity to establish commodity production in Jamaica for internal consumption. Prior to the 1770s, most of the livestock that was used in Jamaica was imported from the North American colonies. By suggesting that animal pens be established on Jamaica, Long was offering a spatial solution to one facet of the disruptive crisis experienced by Jamaican planters in the 1770s. By becoming more self-sufficient in the production of hoofed commodities, the Jamaican planters would be free from the vagaries of trade with the mainland colonies of North America. In a similar light, he suggested that the Jamaica Assembly could encourage the small-scale production of palm oil, saving the planters the expense of importing this commodity and encouraging the development of a middling producing class (Long, 1970 [1774]:414–415).

To encourage the development of small-scale animal pens, palm oil production, and coffee and indigo plantations, Long suggested a scheme that would, at public expense, establish a number of townships in the interior districts of Jamaica. The majority of the settlers in these townships would be small-scale planters, augmented by the artisans and professionals who would be needed to maintain such settlements.

The artisans were to be introduced to save the settlers the time and expense of traveling to acquire necessary tools and articles that could be manufactured locally; a surgeon would be paid from the public treasury until the time came when he would no longer be necessary, or else the township itself could support the surgeon (Long, 1970 [1774]:431). He suggested that each township be supplied with a carpenter/joiner, mason/bricklayer, wheelwright, sawyer, blacksmith, saddler, tailor, and shoemaker (Long, 1970 [1774]:442). The creation of such townships would civilize what was perceived as a dangerous, unproductive wilderness.

Long believed that these townships should be surveyed at public expense, the land taken by eminent domain should the owners be found in default of the terms of their patents, which universally stipulated that any patented land had to be brought into productive use after a certain number of years. Long recognized that much of Jamaica's interior had been patented to sugar planters, who held legal title to the land, but in the case of much of the interior, including the Blue Mountain region, never settled or developed. Much of Jamaica was characterized by a paper cognitive landscape of patents and titles, but a material space of forest and mountain, rarely if ever visited by the absentee holders of the patents. Long hoped to resolve the contradiction between the cognitive and material spaces by creating and settling townships. These townships were to be subdivided into lots of equal size; as the lands were to be seized by the Crown, they could be redistributed (regranted) to new settlers. Significantly, under Long's plan, no settler would be allowed access to more than one lot, ensuring spatial equity among the new middle class. Long based these idealizations on several township schemes that had been undertaken in the North American colonies, particularly South Carolina. He believed, however, that because of the richness and potential productivity of Jamaican land, Jamaican schemes should be actualized on a smaller scale than on the mainland. His feared that portioning out too much land to the proposed middle class could perpetuate the cycle of nonproductivity of Jamaican space (Long, 1970 [1774]:421).

Long illustrated his scheme with an idealized plan of his proposed township (see Figure 12). To promote equality within the white middle class, Long proposed a symmetrical distribution of houses and lots. Order was a key to the class discipline that would be required to make such a scheme successful. Long suggested that all of the houses should be identical, again emphasizing the creation of order and regularity among the middle class: "The houses should be built after one certain

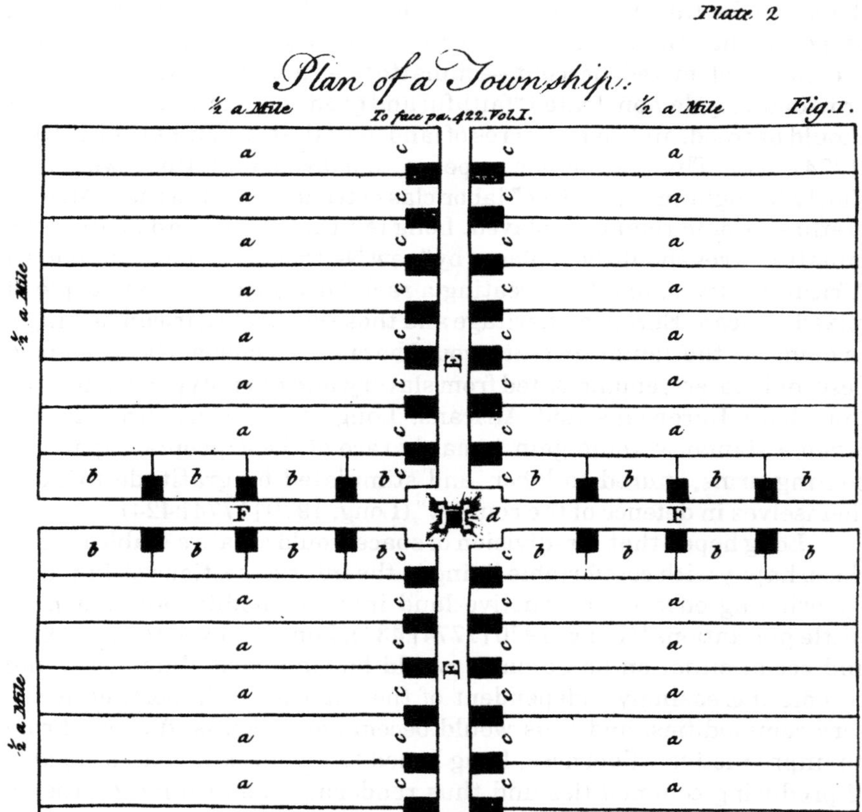

Figure 12. Long's scheme for a Jamaican township. Reprinted by permission from *The History of Jamaica by* Edward Long published by Frank Cass & Company, 900 Eastern Avenue, Ilford, Essex, England. Copyright Frank Cass & Co Ltd.

model, to be approved of by the legislature, and at a certain expense" (Long, 1970 [1774]:421).

Long's scheme in no way intended to replace the enslaved laboring class with a wage-earning working class. He did not call for an end to slavery; thus, the coercive nature of labor extraction would still be in place. Indeed, Long explicitly detailed a scheme by which the emerging middle class would receive subsidized slave labor (Long, 1970 [1774]:423–424). This subsidized slavery would have several effects. It would provide one laborer for each family, thus creating a class of small-landholding, slave-owning agrarian producers and would also set the stage for the creation of a small-landholding mulatto class. Long proposed that the slaves provided by the state to these landholders should be of mixed descent — mulattoes — and that after 5 years, provided they demonstrated "faithful and good behavior," these workers should be freed, and given 5 acres of land near the townships (Long, 1970 [1774]:424). Thus, at public expense, enslaved mulattoes would be slowly brought into the "free" labor class structure, while their African brethren would remain enslaved. Long felt that once emancipated, the mulatto slaves should be replaced by "new Negroes," i.e., newly imported African captives, in effect creating a racial division between people of mixed African–European heritage and those of purely African heritage. In doing so, the Jamaican state would be creating a loyal class of mixed descent laborers, emancipated from slavery, and phenotypically distinct from both Europeans and Africans. Long stated that through this scheme, Jamaica would gain "a hardy race of these people, capable of bearing arms, inured to labor, and stimulated by gratitude to exert themselves in defence of the country" (Long, 1970 [1774]:424).

Long hoped that a redivision of space would create a stable middle class, happy with comfortable living in the interior sections of Jamaica. By bringing once nonproductive land into commodity production on "little plantations" (Long, 1970 [1774]:431), Long believed the economic and social situation on Jamaica would be stabilized; Jamaica would become increasingly independent of the vagaries of importing necessary commodities, and thus would become more self-sufficient. From an unproductive wilderness, Long hoped to invent new spaces capable of producing commodities and thus rendering Jamaica more productive, profitable, and socially stable. This required the manipulation of the class structure by limiting land monopolies, redefining the nature of wealth from landholding to commodity production, and creating a middle class supported by what he felt was reasonable access to produced and producing space. The entire scheme would be reinforced by newly defined spatial relationships.

As discussed in Chapter 3, the planter-physician Dr. David Collins viewed the crisis in the West Indies from a more localized perspective than did Long. Collins's understanding of the crisis focused on the demographic instability of the African populations on plantations. He recognized that a significant relationship existed between the quality of the space in which people lived and the quality of their health. He cited as evidence for this relationship the observation that most slaves who reported to his plantation hospital did so in the morning; he concluded that "next to hard labor and scant feeding, nothing contributes more to the disordering of negroes than bad lodgings" (Collins, 1971 [1811]:115–116).

In order to ameliorate the demographic trends that contributed to the dilemma of the social reproduction of labor, Collins suggested that the West Indian planter class reconstruct the social relationship that existed between the laborers and their domestic space. He asserted that unlike free laborers, the majority of plantation slaves did not have the time to repair their houses when they required maintenance; the houses thus became drafty and damp, inviting disease (Collins, 1971 [1811]:116–117). In response to the health crisis, Collins suggested that the laborers' houses should be well built and situated in such a way as to ensure that the interiors be kept dry and healthy. To this end, Collins recommended that both the interior and exterior walls of the houses be constructed of wattle and daub, rather than of thatch; the floors be raised 6 or 8 inches above the ground and provided with sufficient drainage to ensure that the interiors would be kept dry; and that workers be furnished with wooden bedsteads to "prevent their flesh from being annoyed" (Collins, 1971 [1811]:121). Alternatively, Collins suggested that the planters consider constructing houses of stone, to prevent fires or hurricanes from ravaging the villages and jeopardizing the safety of the working population. He recommended that such stone houses be built as tenements, with three units under one roof, each unit consisting of two rooms. Collins referred to these, respectively, as the "outer one, or hall, and the sleeping room" (Collins, 1971 [1811]:122). He recommended that the hall measure 12 feet square, while the sleeping room be "about ten by twelve in the clear" (Collins, 1971 [1811]:123); as these buildings would be one room deep, by his reckoning, each of the stone tenements would measure 66 feet in length. Collins recognized that the cost of constructing such buildings might be prohibitively expensive for most planters; nevertheless, he suggested that each plantation should have at least one such "hurricane-house" in which the workers could seek shelter during violent storms; the roofs of these houses should be low, and hipped (Collins, 1971 [1811]:124).

Collins, who was a sugar planter in St. Vincent, recognized the relationship between social space and labor discipline. His understanding of this relationship is well expressed in the definition of "discipline" he provides in his treatise: "This term [discipline] embraces two meanings, the one, more comprehensive, includes the rules which direct the conduct of one in subjection to another, the other, the punishment annexed to the breach of these rules" (Collins, 1971 [1811]:169–170). Collins understood that discipline was a means to an end, that end being the most effective extraction of labor from the enslaved workers. He further recognized the need for both punishment and reward, although he warned the planters that overindulgence could encourage the "principles of the slaves" to become "profligate, and their adherence to the interests of their owners more loose" (Collins, 1971 [1811]:171). The proper use of authority, argued Collins, was critical for the extraction of labor (Collins, 1971 [1811]:172). Thus, Collins advised that physical punishment, e.g., whipping or administering "stripes," while sometimes necessary, was not the most effective means of creating a disciplined work force. This logic was not based on compassion, but rather on cold pragmatism. According to Collins, frequent whippings only stimulated the development of thick calluses on people's backs, and thus "destroys their sensibility and renders its further application of little avail" (Collins, 1971 [1811]:173). Furthermore, Collins argued, fear of whipping was the most common reason for the slaves to attempt to flee the plantation. Diminishing the use of the whip would therefore reduce the frequency of attempted escapes.

A far more efficient mode of punishment, he argued, was confinement, that is, constricting the spatiality of movement. For recidivist runaways, those who continually challenged the planter's hegemony over movement through space, Collins suggested that they be assigned a supervisor, who would accompany them to work, and be responsible for overseeing their daily work and movements. On retiring from the field, the laborers should be confined, under guard, in the slave hospital. This constant supervision, this limited access to social space, would eventually lead to the breaking of the slaves' spirit, while securing their labor and preventing them from running away (Collins, 1971 [1811]).

Collins believed that such manipulation of social space would create an orderly and disciplined labor force. He further recognized the efficacy of constant surveillance in his notes about the placement of slave villages in relation to the planters' or overseer's quarters. In his words, the planters should "take care that the [slave] houses be not too distant from the family dwelling, so that the proprietor, or his manager, may at all times have an eye to his gang, to be informed of their proceedings, to permit and encourage innocent mirth, but to suppress turbulent contentions" (Collins, 1971 [1811]:118).

Collins's advice on confinement seems to have been taken up by at least some of the Jamaican planters. Matthew ("Monk") Lewis, the proprietor of several Jamaican sugar estates, is one example of a planter who publicly preferred confinement to lashing. Born in 1775, Lewis was a planter by default, having inherited his estates at his father's death in 1812. According to the brief biography written by Mona Wilson (1929), Lewis was acquainted with a number of important literary characters, including Goethe, Byron, Shelley, Coleridge, and Walter Scott. He was for a time attached to the British Embassy at the Hague and served from 1796 to 1802 as a Member of Parliament, but his literary aspirations seem to have outstripped his political ambition. He wrote several novels, his most famous, *The Monk*, was published in the mid-1790s. Lewis traveled twice to Jamaica to inspect his estates, particularly the condition of the slaves. On his second trip he contracted yellow fever, dying in 1818 during his return voyage to England (Wilson, 1929).

Influenced by the Romantic movement, Lewis expressed humanitarian concern for the welfare of his slaves; such sentiment did not, however, induce him to emancipate them. He did travel to Jamaica in 1815 and again in 1817, ostensibly to investigate the well-being of his laborers. In the account of his journeys, first published in 1834, Lewis echoed Collins's earlier advice about the efficacy of confinement as a mode of discipline. In disciplining a slave named Toby, who refused to work during several days of the crop season, Lewis suggested that "Toby did not mind three straws" about the six lashes that were stroked across his back. Determined that Toby's mutiny would not go unpunished, Lewis had the recalcitrant worker locked into a small room in the hospital, where he was held for 3 days, over the Easter holiday weekend. Lewis asserted that this punishment was far more effective that the stripes (Lewis, 1929 [1834]:316). Lewis suggested that such forced seclusion from society should be the punishment for those who attempt to feign illness to avoid work. Once locked in the hospital, these laborers would only be allowed to leave when they declared themselves fit for work. Lewis reported that this measure was quite effective in curbing this form of resistance; after adopting this measure, he reported, recidivism was nearly abolished, and the number of hospital patients had decreased by more than half (Lewis, 1929 [1834]:317).

IMAGINING COFFEE PLANTATIONS: LABORIE'S SCHEME

Pierre-Joseph Laborie was a French planter from St. Domingue harbored in Jamaica during the Haitian revolution. He was a lawyer born in St. Domingue and served on the administrative council of that colony while it was occupied by the British during the French Revolution. Although he had retired from law to an estate in the highlands of St. Domingue, during the uprisings of the 1790s he sought refuge in Jamaica, where he bought a coffee estate (Higman, 1988:159). In gratitude for the hospitality he was afforded in Jamaica, he took it on himself to educate the British planters of Jamaica on the cultivation of coffee. This enterprise resulted in the publication of an instructive treatise on coffee production titled *The Coffee Planter of Saint Domingo*, published in 1798. This book was an instructive guide to every aspect of coffee cultivation, from the original creation of the coffee plantation as productive space, through the processes of cultivation and preparing crops for export. Higman argues that this was the most influential manual consulted by Jamaican coffee planters (Higman, 1988:159). An analysis of Laborie's manual allows us to interpret how the spaces of coffee plantations were imagined by potential planters prior to their being actualized on the material landscape of the island.

Imagining the Landscapes of Production: Plantation Design

Laborie's treatise included an idealized plan for the proper way to design and construct the landscape of production on a coffee plantation. He illustrated this with a schematic representation of the ideal plantation layout (Figure 13). His idealization imagined coffee plantations laid out on a symmetrical grid plan, in which a set of carefully ordered coffee fields were spatially balanced against a series of cultivated wood lots. Laborie argued that the domestic and industrial spaces (which he collectively defined as "the houses and settlements") should be placed in the geographic center of the plantation. Pastures and provision grounds should be located near the settlements; in his representation, these are located to the south of the central area. To the west of the domestic/industrial area should be a savanna, or pasture, and to the east "Negroe grounds." Notably, the Negro grounds are divided on Laborie's graphic representation into a system of small agricultural fields, suggesting that these were spaces cultivated by the enslaved for their own provisions.

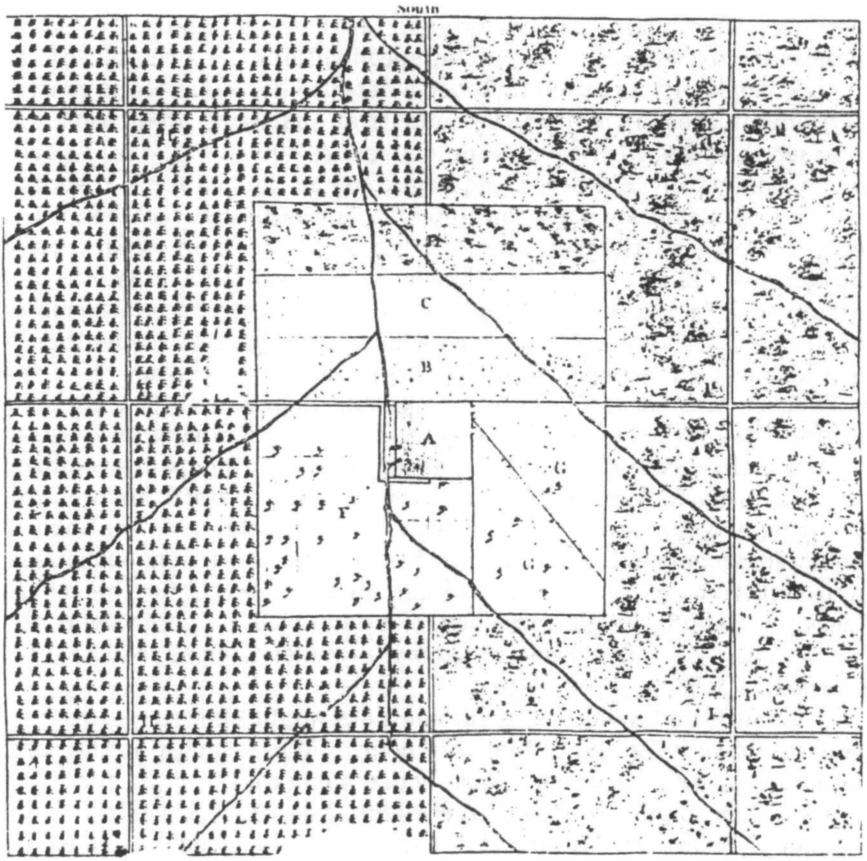

Figure 13. Laborie's scheme for the ideal coffee plantation layout. Source: Laborie, 1798:plate 3.

In his cognized plantation layout, Laborie suggested that coffee fields be symmetrically placed in the areas surrounding the central settlement zone. Where the topography allowed, the fields should be of approximately equal size, connected to each other and to the industrial complex by a series of straight roads. Laborie measured his symmetrically designed fields by a unit called "a square of land," equivalent to just under 3 acres. A square that had optimal soil quality, topography, and sunlight exposure would contain 13,611 trees, each occupying 3 square feet. The distance between the trees would vary depending on the quality of these variables; Laborie suggested that as few as 932 trees could be planted in a square. Laborie recognized, however, that because coffee plantations would be located in highlands, it could be impossible to conform exactly to his specifications (Laborie, 1798:113).

Laborie's idealization included a detailed depiction of the central settlement area, including the industrial spaces of production. He advised that this entire area should measure 400 feet square, an area of approximately 3⅔ acres. He argued that the planter's dwelling house and coffee store be contained in one building, located in the center of the settlement area. In his idealized plan, this central building would be oriented to the south, overlooking a geometrically arranged formal garden, flanked on one side by drying platforms or barbecues. These platforms were where the harvested coffee beans would be dried between pulping and grinding. To the north of the barbecues was the pulping mill complex, including the cisterns required to separate the coffee pulp from the beans during processing. To the right of the garden, Laborie represented the spaces required for the provisioning and livestock of the estate staff: an orchard, poultry yard, and stables. To the right of the planter's house, Laborie depicted dependencies that included the planter's kitchen, outhouses, and a slave hospital, with a yard behind the hospital for the convalescents. Behind the planter's house, hidden by a hedgerow, were the slave houses, represented in Laborie's model as long rectangular barracks (Laborie, 1798:plate 4).

Laborie conceded that some planters might choose to construct separate dwellings and coffee store houses. When such was the case, Laborie suggested that the store house measure 60 by 34 feet. The ground floor should contain three rooms, each 20 feet square, and a 14-foot-wide gallery running the length of the building. Coffee would be stored in the three rooms and the garret, while the sifting, culling, and weighing of coffee would take place in the long gallery. He asserted that the building should be well ventilated and fenestrated to ensure that the coffee would not molder, and "must be of mason's, or the best carpenter's work." He cautioned that if located on the south side of the barbecues, the planter should be sure that the building did not cast a shadow over the barbecues, which would prohibit the proper drying of the pulped beans (Laborie, 1798:87). For those planters who chose to separate their living quarters from the coffee store, Laborie made no specific suggestions on the design of dwelling houses, save that they should contain a chimney and glass windows, as coffee plantations were likely to be located in a cool and rainy climate (Laborie, 1798:96). For the planter who chose to "attend chiefly to his manufacture" (Laborie, 1798:96), Laborie suggested that the planter's dwelling space be located in the same building as the coffee works and store. He recommended this arrangement for the planters, as he felt that workers engaged in the final stages of coffee production required "constant watching" (Laborie, 1798:97). By locating the coffee works in the same building as the planter's domestic space, both the planter and the workers would be aware of the increased level of surveillance over the tedious work of preparing coffee for export. Laborie recommended that the planters take measures to ensure that if they lived in the same building as the coffee store, they be isolated from the noise of the machinery, and the tobacco smoke of the workers (Laborie, 1798:96).

Laborie commented on what he felt was the ideal spatial organization for the housing of the enslaved workers on a coffee plantation. According to his idealization, the workers should live in a type of tenement that he defined as a "barracks." He argued that such barracks should measure no more than 100 feet if roofed with shingles or boards, or 80 feet if they were thatched. Within this large structure, 200 square feet should be allocated to "every two negroes"; however, he does not mention whether these pairings should be conjugal (Laborie, 1798:95). The spaces so allocated should be divided into two rooms, one for cooking and the other for sleeping; a gallery or leanto could be added to the back of the structure to serve as a chicken coop for each pair of workers. These buildings, he argued, should be earthfast, i.e., hardwood posts would be sunk into the ground to provide the standing members for these structures. The walls should be made of wattle and daub and in order to minimize the risk of fire, each apartment should be provided with a masonry hearth. The planters, warned Laborie, should play an active role in designing and actualizing these domestic spaces, and should never "suffer the negroes to make their own huts" (Laborie, 1798:95–96).

Laborie suggested to the Jamaican planters that designating provision grounds should be among the first things done on a new coffee plantation. He argued that access to provision grounds served to attach workers to an estate, enabling them "to reap comfort from their own industry" (Laborie, 1798:38). He asserted that the grounds be located together, rather than dispersed, and should be divided into lots of "twenty-five paces square, allowing sixteen negroes to a square of land" (Laborie, 1798:38). He maintained that the boundaries of each lot should be demarcated by small hedgerows of peas; half of each plot should be cultivated in plantains, the other half left to the discretion of the worker cultivating the plot. Maintaining that such symmetry would create social order, he emphasized the need to create regular, symmetrical plots (Laborie, 1798:39).

Each of these various schemes, proposed prior to the abolition of slavery, had an affect on the discourse concerning the design and construction of coffee plantations in the early nineteenth century. This was the first stage in the spatial reorganization of Jamaica resulting from the change in the global mode of production. Another phase of restructuring resulted from the basic change in the relationship between the planters and workers that came with the end of slavery. Just as plantation theorists proposed new cognitive spaces during the pre-emancipation years, a new generation of social engineers suggested ways by which space could be used to ensure that the basic class and racial structures that had been in place during slavery would be maintained. An analysis of these cognitive spaces follows.

THEORIES ON FREEDOM: IMAGINING POSTSLAVERY JAMAICA

The institution of British West Indian slavery was abolished by a legislative act of Parliament passed in 1833 and enacted between 1834 and 1840, known as the Emancipation Act. The terms of Jamaican emancipation were thus carefully calculated before slavery was ended, a situation unlike the abolition of slavery in Haiti or Spanish South America, where slavery was ended through revolution, or in the United States, where abolition was a result of a violent civil war (Green, 1976:99ff). Because abolition in the British West Indies was debated and eventually legislated, the intentions of those interested in accomplishing this end can be interpreted. For the purpose of this study, I will present an analysis of the major imagined spatialities that emerged during the emancipation debate. The historian Thomas Holt maintains that of the many plans for the implementation of emancipation, two — by Henry Taylor and Henry George Grey, respectively — are particularly useful in "exposing the fundamental concepts shaping emancipation" (Holt, 1992:42).

As British emancipation was a legislative act, the responsibility of much of the design fell to career bureaucrats working for the British government. Henry Taylor was such a career bureaucrat; by 1833, at the age of 33, Taylor had already been working as a clerk in the Colonial Office for a decade (Holt, 1992:42), having used personal connections to establish himself in that office in January 1823 (Taylor, 1885:63–64). A part-time poet and member of the London literary set, Taylor turned down several opportunities for promotion, including the governorship of Upper Canada and the office of Permanent Undersecretary for the

Colonial Office, choosing to remain a part-time bureaucrat with literary aspirations (Holt, 1992:43). As a clerk in the Colonial Office, Taylor was acquainted with most of the leading members of the antislavery movement, including Henry George Grey (Viscount Howick and later the third Earl Grey), William Wilberforce, and Wilberforce's nephew James Stephen (Taylor, 1885:123–124).

Taylor presented a plan for emancipation to his superiors in January 1833 (Holt, 1992:43). He believed that the contradictions in spatial understandings between the planters and the enslaved were obstacles to transforming the enslaved population into a wage-labor work force. These contradictions were particularly acute in Jamaica. Taylor argued that the intersection of several spatial phenomena in Jamaica created a particularly difficult situation for the implementation of an emancipation measure based on the ideology of liberal capitalism. These spatialities were characterized not only by an island-wide low population and a vast supply of uninhabited and undeveloped lands in the interior, but also by the independent subsistence economy of the enslaved Africans. This subsistence economy was reinforced by a spatial understanding of provision grounds as land in possession of the enslaved, and treated by the masters as the virtual private property of the enslaved. Taylor commented that customarily, the enslaved population of Jamaica had provided for its own subsistence by working one day a week in their provision grounds. The key to making emancipated labor successful in the maintenance of Jamaica's export economy, Taylor argued, was to develop a scheme that would encourage the workers to sell their labor to the estates (Holt, 1992:43).

In Taylor's opinion, once emancipated, the workers needed to be endowed with an internalized work discipline; he did not believe that an unlimited desire to purchase commodities would be sufficient motivation to spontaneously encourage the freed workers to remain on the estates (Holt, 1992:43–44). Without implementing carefully calculated measures, the population would scatter throughout the island. Taylor argued that the population needed to be "condensed" in order to be "civilized," i.e., to provide labor for the planters. If they were not regulated, the population would become, in Taylor's words, "squatters and idlers . . . living like beasts in the woods" (Holt, 1992:45). In order to implement the discipline of wage labor, Taylor argued that the slaves should be responsible for purchasing their own freedom. To this end, the British government would purchase one day's worth of freedom from the planter for each slave. In this one free day per week, the laborer would be allowed to sell his or her labor power in the labor market, and, with thrift, would be allowed to purchase additional days of freedom, until the remaining six days were bought and the person was fully free (Holt, 1992:46). In doing so, the laborers would be gradually introduced to wage labor, one day at a time as it were, and gradually internalize the emerging social relations of production mediated by the sale of labor power. Taylor believed that such a scheme would redefine the spatialities of Jamaica, and would encourage the moral order of competitive capitalism.

Taylor's plan, written as a detailed memorandum, was never adopted as policy; in fact, it never made it out of the Colonial Office (Holt, 1992:47). His superior, Henry George Grey, rejected this scheme in favor of one of his own, which was sent to the Cabinet in place of Taylor's. According to Taylor's autobiography (1885), as early as 1831, during the fallout from the Baptist War, Grey had developed his own proposal for emancipation. Although Taylor's Colonial Office colleagues, Grey and Stephen, agreed that the responsibility for drafting a plan should fall to Taylor, Grey expressed his own ideas to Taylor, particularly concerning the redefinition of the relationship the workers had to the productive space of the provision grounds. According to Taylor's account, Grey believed that strategy implemented to instill labor discipline should be based on a strict control over space.

The plan that Grey eventually forwarded to the Cabinet called for restrictive measures on the spatiality of the colonies. The implementation of vagrancy laws and the assessment of a land tax, according to Grey, would serve the dual purposes of restricting the unauthorized use of unoccupied lands (which in the colonial world of Jamaica in the 1830s were ostensibly owned by the British government) and compelling small farmers to sell their labor in order to raise the capital to pay the land tax (Holt, 1992:48; Taylor, 1885:125). Grey's plan was never implemented, as it was rejected by the Cabinet based, in part, on the objections from Lord Mulgrave who at that time was governor of Jamaica (Holt, 1992:48). Several weeks after the defeat of his policy proposal, Grey resigned his post as undersecretary in the Colonial Office (Holt, 1992:48).

Emancipation legislation was finally introduced in Parliament in May 1833, under the direction of the new secretary of the Colonial Office, Edward Stanley. It was Stanley's legislation, drafted by Taylor's colleague James Stephen, that eventually was adopted as the Emancipation Act in December 1833 (PP, 1835:50, 273). The primary tenets of the Act as approved and implemented held that slavery was to be abolished on August 1, 1834, and replaced with a temporary labor system, known as "apprenticeship," which itself was slated to be abolished and replaced with a free wage labor system by 1840. The West

Indian planters were to be compensated for a percentage of the value of their slaves; to this end, Parliament approved an allocation of £20,000,000, a fantastic sum for the period.

The apprenticeship system was designed as a strategy through which the formerly enslaved people would be introduced to the discipline of wage labor. Under this act, the formerly enslaved laborers were divided into three classes of apprentices: (1) praedial-attached laborers were people usually employed in agricultural pursuits "upon lands belonging to their owners"; (2) praedial-unattached apprentices were those engaged in agricultural pursuits "upon lands not belonging to their owners"; (3) nonpraedial apprentices included all others, including domestics (PP, 1835:50, 273). For the length of their apprenticeship, each apprentice would be compelled to work for their former masters for no more than 45 hours per week. The remainder of their time could be spent working in their provision grounds or selling their labor in the market, including to their former masters, for wages.

Although no longer legally classified as slaves, the spatialities of the apprentices under the Emancipation Act were severely restricted. During the period of the apprenticeship, laborers were required to gain permission from their planter supervisors in order to leave the plantation to which they were attached. The Emancipation Act provided a scale of punishments for those caught violating this restriction of the spatiality of movement. An apprenticed worker absent without leave for any 2 days within a fortnight, or any 2 consecutive days, was to be defined as a "deserter." The punishment for desertion was either 1 week in the "house of correction" (the workhouse) or a flogging of 20 "stripes." An absence of 3 or more days earned a worker the title "vagabond" and was punishable by 2 weeks in the workhouse or 30 stripes. Any apprentice caught wandering off the property of their "employer without a written permission" was liable to arrest and punishment (PP, 1835:50, 277). Any apprentice caught trying to flee Jamaica entirely would be sentenced to up to 6 months in the workhouse or flogged with up to 50 stripes (PP, 1835:50, 278).

Although permitting the laborers to remain in possession of their houses and grounds, the Emancipation Act provided the planters with the right to inspect and control the houses and provision grounds of the apprenticed employees, thereby giving the planters the right to invade the domestic space of the workers at will. If a planter concluded that an apprentice allowed his or her "house or provision ground to fall into decay and bad order," the planter could intervene, and restructure the domestic or agricultural space of the workers as he saw fit. In doing so, the apprenticed worker would be docked the number of hours the planter deemed necessary to reconstruct the space, but not to exceed 15 hours in a given week (PP, 1835:50, 278). This measure allowed a significant amount of control to the planters to define the domestic spaces and the spaces of production in which the workers lived and operated.

During the transition to freedom, as imagined by the Emancipation Act, the planter class would use these controls over the spatiality of the work force as a means to control labor. Free movement would continue to be restricted; thus, the planters would retain their hegemony over the spatiality of movement. While the laborers would continue to provide their own provisions and retain access to those spaces they had occupied during slavery, the planters were to have the right to intervene and restructure any domestic or agricultural spaces that they defined as inappropriate to the existing social order.

As Holt argues, the variations on social engineering debated among British politicians and bureaucrats, while differing in many respects, shared certain elements. While hoping to maintain a racially divided society in which whites maintained social superiority over blacks, each of these schemes rejected the most extreme racist position that Africans were naturally indolent and inherently inferior to Europeans. Those in favor of emancipation believed that freed slaves could, with the proper training, become disciplined members of a wage labor force. Each of these schemes was grounded in the ideology of competitive capitalism, which held that constraints on the expression of self-interested behavior must be removed in order for the political economy to operate most effectively. Holt further argues, and I must agree, that each of the theorists on emancipation believed that slavery was incompatible with nineteenth-century capitalism, as free laborers were more efficient and profitable than enslaved laborers (Holt, 1992:50).

CONCLUSION

Proposing social and spatial restructuring in the years leading to the end of slavery can be interpreted as the negotiation of cognitive spaces. Each of the plantation theorists interested in rationalizing plantation space, as well as the politicians concerned with restructuring the social space of Jamaica following the end of slavery, contributed to a spatial discourse on how best to reorganize the social landscape of Jamaica. Each of these authors had as his goal the restructuring of the social order; each recognized that redesigning the relationships people had with space was key to any such social engineering. Because these were interpretations of social and material space, they fall under the category of cognitive space. In the following two chapters, I analyze how the imagined, cognitive spaces were manifested in the social and material spaces of the Yallahs region. Chapter 6 considers how these spaces were negotiated in the decades prior to emancipation; Chapter 7 analyzes postemancipation developments.

6

The Spatialities of Coffee Plantations in the Yallahs River Drainage 1790–1834

A slave, being a dependent agent, must necessarily move by the will of another, which is incessantly exerted to control his own; hence the necessity of terror to coerce his obedience.
— Dr. David Collins, 1811

INTRODUCTION

During the eighteenth century, the conditions of life for the great majority of people in Jamaica, both enslaved and free, were influenced by the production and distribution of sugar. As was the case in most British colonies that relied on monocrop production, fluctuations in world markets for this commodity affected those involved in its production. In the closing decades of the eighteenth century, changes in the global logic of capitalism began to effect changes in the political economy and social landscape of Jamaica. These were decades of crisis for the plantation economy of Jamaica. Among the interesting socioeconomic phenomena that resulted from this crisis was the rapid florescence — and equally rapid abandonment — of large-scale, estate-based coffee production. This florescence was marked by a brief coffee boom that began around the turn of the century, peaked in the first decades of the century, and collapsed in the late 1830s. By the 1850s, most of Jamaica's coffee plantations had been abandoned or transformed into some other spatial entity (Higman, 1988). The intensity and brevity of the Jamaica coffee boom produced an interesting, tightly datable data set. By examining the documentary and archaeological records of Yallahs region coffee plantations, it is possible to analyze how specific new spatial forms were designed and implemented on the physical landscape and how these forms were used to construct and reinforce new social relations of production. In this chapter I consider how these sociospatial phenomena resulted in the production of the spatial entity known as the coffee plantation in the Yallahs region, and how the social spaces of those plantations shaped the lives of those who lived them.

PRODUCTION OF LANDSCAPES AND SPATIALITIES

Although the design and production of space on Yallahs coffee plantations differed from sugar plantations in the type and number of buildings required, the planters adapted some of the preexisting spatial logic of sugar plantations to the demands of coffee agriculture. Thus, in the early nineteenth century when most of these concerns were new, the estates would have easily been identifiable as plantations by a visitor unfamiliar with the specific spatial layout of a coffee estate. As was the case on sugar plantations, there was a race- and class-based segregation of domestic space incorporating the living areas of the planters, the white estate staff, and the enslaved African laborers. Domestic space was segregated from the space of industrial processing, which in turn was separated from the agricultural fields. The segregation of space on coffee plantations is represented in the idealized model of a coffee plantation rendered by Laborie in 1798 (see Figure 13). In this representation, one can see the separation of activity zones on a plantation; the domestic area is located at the symbolic center of the plantation, flanked by woodlands and coffee fields. The provision ground for the estate staff is detached from both the domestic space and the provision grounds of the workers.

Thus, on a basic level, the material space of a coffee plantation can be conceptualized as having been divided into several spheres: the domestic space of the elite comprised of the great house, overseer's house, and their dependencies; the domestic space of the enslaved comprised of the slave villages; the agricultural space of elite commodity production comprised of coffee fields and pastures for draft animals; the agricultural space of food and commodity production of the enslaved comprised of provision grounds; the industrial space of production comprised of the mill complex; and what we might define as intermediate space, that is, the areas of the plantations that were neither under production of coffee nor provisions and which were sometimes defined as "ruinate" or forest.

To best understand how these particular spaces influenced the negotiation of social relations between the enslaved and the elites, it is necessary to lay out how these material spaces were manifested on the landscape. To accomplish this, I will first present the evidence that exists for how these specific spaces were constructed. I will then consider how social spaces on coffee plantations in the Yallahs region were negotiated by discussing how the social relations of production were actualized in these material spaces. Because the spatialities of the plantations are best reconstructed by interpreting the process of actualizing social space on the material landscape, however, a discussion of the material spaces must come first.

Producing the Landscapes of Production

In analyzing how spaces and landscapes were produced in the Yallahs region, one must begin by looking at how the preproductive space of the region was reconceptualized as a space of production. As may be true of any region that is re-created into a space for commodity production, this process began with the division of space into segments owned by individual proprietors, i.e., the landscape was redefined into "property." The earliest known record of this process for the Yallahs region is an undated late eighteenth-century rendition of property lines for several estates (Figure 14). The map portrays a series of estate boundaries. The only landscape features it depicts are the Yallahs River and several of its tributaries, the location of several houses, a waterfall (cascade), and a bridge over the Fall River.

Significantly, the estates that are outlined on the map include the name of the patentees of the land and the date when the land was first patented by an individual planter/speculator and/or the land was first surveyed. This evidence suggests that some of the land in the Yallahs region was patented as early as 1683; the majority of the estates depicted, however, were surveyed between 1746 and 1773, suggesting that in the third quarter of the eighteenth century, the landscape of the Yallahs region began to enter into the consciousness of Jamaican planters as a potential space of commodity production. Because these lands were not identified as plantations, it is probable that although patented by the Crown to land speculators, most of these parcels were not yet developed. By the time this map was drawn, however, several houses were located in the region, including great houses later associated with Mt. Charles and Mavis Bank. It seems likely that, at least on these estates, white planters had begun the process of clearing the forests and establishing homesteads as a prelude to the creation of coffee plantations.

This map portrays the estates as vacant lands. There was no attempt to depict any industrial works associated with coffee produc-

Figure 14. Earliest known rendition of estates in the Yallahs region. Courtesy of the National Library of Jamaica.

tion, nor was any spatial division within the individual parcels noted. This image illustrates the bounding of once "unproductive" or wild lands into a spatial form designed for capitalist commodity production. It is likely that this image depicts a cognized space representing the earliest stage of the production of plantation space: the actual definition of the boundaries between productive spaces, and the delineation and bounding of space as private property. Significantly, at this point in time the estates are identified on the map by the name of the owner. In later depictions, estates are nearly always identified by the name of the plantation (e.g., "Mavis Bank," "Chesterfield") as well as by the name of the owner or estate manager. This development reflects a transition in the cognitive space of these lands, as vacant properties were transformed into production and spatially defined as plantations.

Of the eight properties whose acreages are listed on this map, six are listed as being comprised of 300 acres. One includes 500 acres and the other 40 acres. This suggests that the lands patented in the interior were carved into relatively moderate estates by Jamaican standards, less than a third of the size of the 1045 acre average for eighteenth-century Jamaican estates calculated by Richard Sheridan (Higman, 1988:16; Sheridan, 1974:219). The relatively small size of the patents suggests that from their inception as spatial entities, coffee plantations were constructed as smaller agricultural concerns than sugar plantations.

An analysis of a map of the confluence of the Yallahs and Clyde Rivers, dated 1801, reveals a similar set of processes. This map (STA 247), drawn by the surveyors John Murdoch and Patrick Hewitt Keefe, also depicts the earliest stage of plantation construction, but for an area upriver and inland from the area depicted in Figure 14. As was the case with the earlier map, this image depicts the outline of a number of estates, specifically the boundary of a parcel of land conveyed from Dr. Colin McClarty to Dr. Alexander McClarty; this latter parcel is identified as Clydesdale plantation. The explanatory text accompanying the image defines the bounded areas as "runs of land," not as plantations or estates, again suggesting that these were properties not yet defined as distinct spatial units of production. The only exceptions to this are the run of land from which the McClarty piece was conveyed, which is defined as Chester Vale plantation, Clydesdale itself, and another parcel conveyed from Colin McClarty to William Griffiths identified as "Newport Hill plantation." Significantly, all of these properties were once part of a larger estate, Chester Vale, owned by Colin McClarty. Thus, the map depicts the first stage in what would be the creation of two new coffee plantations through the partition of a larger estate.

All of the bounded properties surrounding Newport Hill, Chester Vale, and Clydesdale are depicted as vacant lands, identified only by the name of the proprietor and the acreage contained within the property. This again suggests that the majority of these properties were not yet involved in coffee production. The parcel that became Clydesdale includes a sketchy rendition of what is most likely an industrial works; a later map (STA 241) corroborates this as the location of the coffee works. Thus, when Colin McClarty conveyed the land to Alexander McClarty, he may have conveyed a parcel of land already in the process of being materially defined as a coffee plantation.

As was the case with the earliest map of the region, the majority of the properties identified on STA 247 contain approximately 300 acres. Parcels attributed to Robert Jackson, John Madder, Henry Ram, and two to Alexander Montier each are identified as containing 300 acres. Following the partition of Chester Vale, the newly created Clydesdale contained 300 acres. Although Newport Hill, conveyed to William Griffiths, contained only 160 acres, McClarty had also conveyed an abutting parcel containing 156 acres to the same Griffiths. Taken together, Griffiths's adjoining properties thus contained 316 acres. Thus, with the exception of the truncated Chester Vale (874 acres after partition) and a 600-acre parcel belonging to Edward Manning, all of the parcels of land surrounding Clydesdale contained approximately 300 acres.

Taken together, these two maps indicate that the first phase in the production of the spatiality of coffee plantations in the Yallahs region entailed the partition of the mountain landscape into private property, containing regular parcels of a standardized size, based on increments of approximately 300 acres. With the exception of Clydesdale and the elite houses depicted on the earliest map, these early renditions of space depict an empty landscape. The division of this empty landscape into culturally constructed bounded units, as depicted on the maps, was the first stage in the process of the production of commodity-producing space.

The Material Space of Coffee Production I: Coffee Fields

The second critical stage in the production of commodity-producing space is the material preparation of land for said production. In the case of the Yallahs region, this process involved the transformation of upland forests into coffee fields. Again, the cartographic record of the region provides a tool to interpret how this transformation occurred. Unfortunately, few preemancipation maps depicting the division of space into coffee fields for the Yallahs region survive. However, an analysis of those few that do survive from the Yallahs region coupled with the analysis of maps from areas surrounding the immediate Yallahs region provide data from which to draw some tentative conclusions about this process.

A plan of Mavis Bank plantation drawn in 1808 by John Pechon and reproduced by Barry Higman (Higman, 1988:168) corroborates that by this date Mavis Bank was in full operation as a coffee plantation. At the time of its survey, 75 of Mavis Bank's 302 acres were divided into six coffee pieces, ranging in size from just over 4, to just over 15 acres. The remainder of the plantation's acreage was divided into parcels dedicated to the production of guinea grass for animal fodder (c. 34 acres), provision grounds probably set aside for the white estate staff (c. 22 acres), "Ruinate Land & Negro Grounds" which may have included the slave village (c. 73 acres), a second parcel of "Negro Grounds" and provision grounds for the enslaved population (c. 79 acres), and a plantain walk, which would be somewhat equivalent to a garden that produced food for the white estate staff (13 acres). Thus, in 1808, approximately 25% of the estate lands were actually dedicated to the production of coffee, while at least 50% of the estate was dedicated either to the cultivation of provisions by the enslaved for their own use, or else was in "ruinate," a condition of land defined as uncultivatable by the elites.

According to Barry Higman (1986), Pechon also rendered a plan of Radnor plantation in 1808. While this map is currently unavailable for analysis, Higman redrew the layout of the estate based on the 1808 map, publishing his rendition in 1986 and again in 1988 (Higman, 1986, 1988). According to his analysis of this plan, the estate contained 689 acres in 1808. At that time, Radnor contained 14 coffee pieces totaling 236 acres, 83 acres were planted in guinea grass, "14 acres in plantains, 133 acres in Negro grounds and provisions, and a further 222 acres in woodland and Negro grounds" (Higman, 1986:84). Thus, 35% of the estate was planted in coffee, while 52% was identified with the enslaved population, or as woodland (Higman, 1986).

A plan of Clydesdale plantation rendered by Frances Ramsey in 1810 exhibits a similar division of space (Figure 15). As was the case with Mavis Bank, Clydesdale was divided into six coffee pieces. However, the coffee fields at Clydesdale tended to be a bit larger, ranging in size from just over 9 to just over 24 acres. According to this plan, 83 of Clydesdale's 329 acres were in coffee; thus, about 25% of the plantation was dedicated to the direct production of the commodity. Only 7 acres were under the cultivation of guinea grass, while 230 acres were

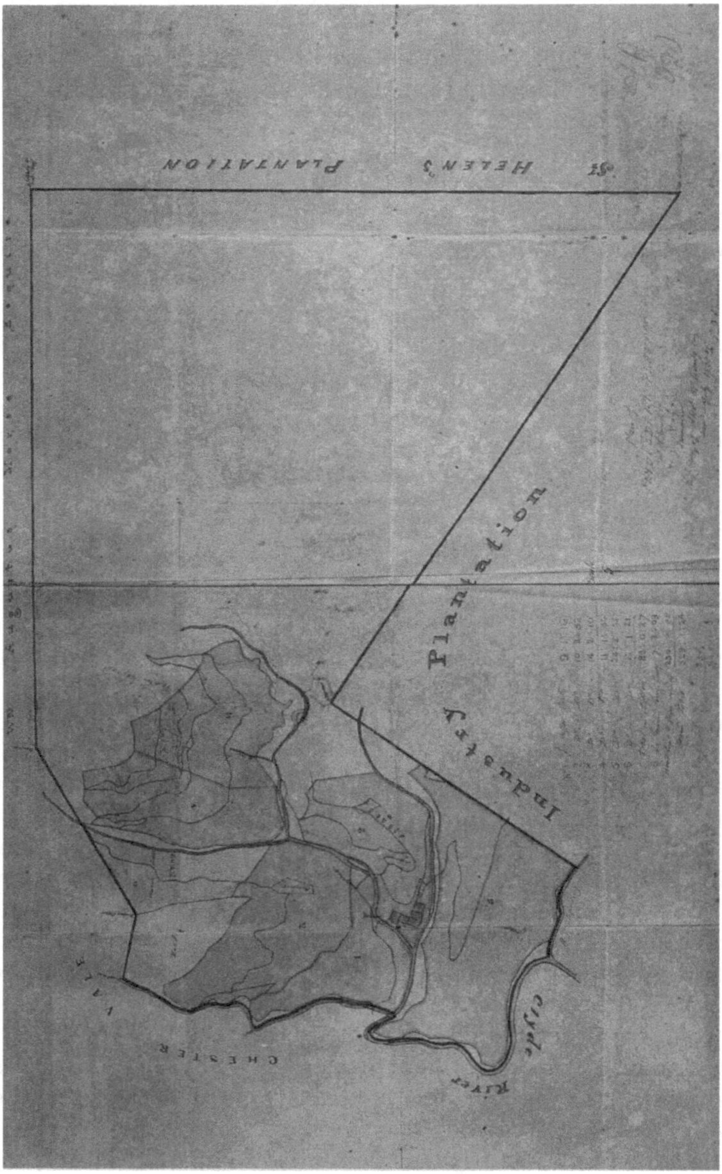

Figure 15. Clydesdale estate plan by Francis Ramsey, 1810. Courtesy of the National Library of Jamaica.

identified as "Woodland & Negro Grounds." In this case, 70% of the estate land was not dedicated to commodity production, and was more readily associated with the enslaved than the white population.

The Material Space of Coffee Production II: Coffee Works

Once harvested, the coffee crop proceeded through several processes prior to its being shipped. These processes required the extraction of coffee beans from ripe berries, the separation of the beans from pulp, and the drying of the beans. This was accomplished with specialized apparatus. The coffee works, i.e., the industrial complex of a coffee plantation, thus would include, minimally, a pulping mill to extract the beans from the berries, a set of drying platforms known as barbecues, and a second machine for removing a thin skin, known as parchment, from the dried beans. The process of pulping required a vast amount of water, not only to power the machinery, but also because the process of separating the beans from pulp required suspending the pulp in water and then separating the beans from the solution. Thus, coffee works also required holding tanks or vats and an aqueduct system to divert water from a river, stream, or spring to the coffee works (Higman, 1988; Laborie, 1798; Montieth, 1991).

While several of the coffee estates in the Yallahs region retain ruins of nineteenth-century plantation works (e.g., Clydesdale, Chesterfield, and Sherwood Forest), others do not (e.g., Whitfield Hall, Radnor, and Arntully). Fortunately, three of the surviving coffee works are located in three subareas within the Yallahs region. Clydesdale is located on the Clyde River, a tributary of the Yallahs, and is found in the far northwest of the study area (Figure 16). During the nineteenth century, it was located in the parish of Port Royal. Chesterfield is located in the broader Yallahs valley proper, is found in the geographic center of the study area, and was located on the boundary between Port Royal and St. David. Sherwood Forest is located near the Negro River, is in the eastern section of the study area, and was located in the parish of St. David. These three plantations also provide good temporal coverage. Chesterfield and Clydesdale provide coverage for the first coffee boom discussed in earlier chapters, as the former was operating as a plantation as early as 1794, and the latter was organized as a plantation in 1800. Sherwood Forest offers information about later transitions in the coffee industry, as it was operating as a plantation at least until 1852. Thus, these three plantations offer a sense of the range of variation of plantations based on their position in material (geographic) and social (parish) space, as well as temporally.

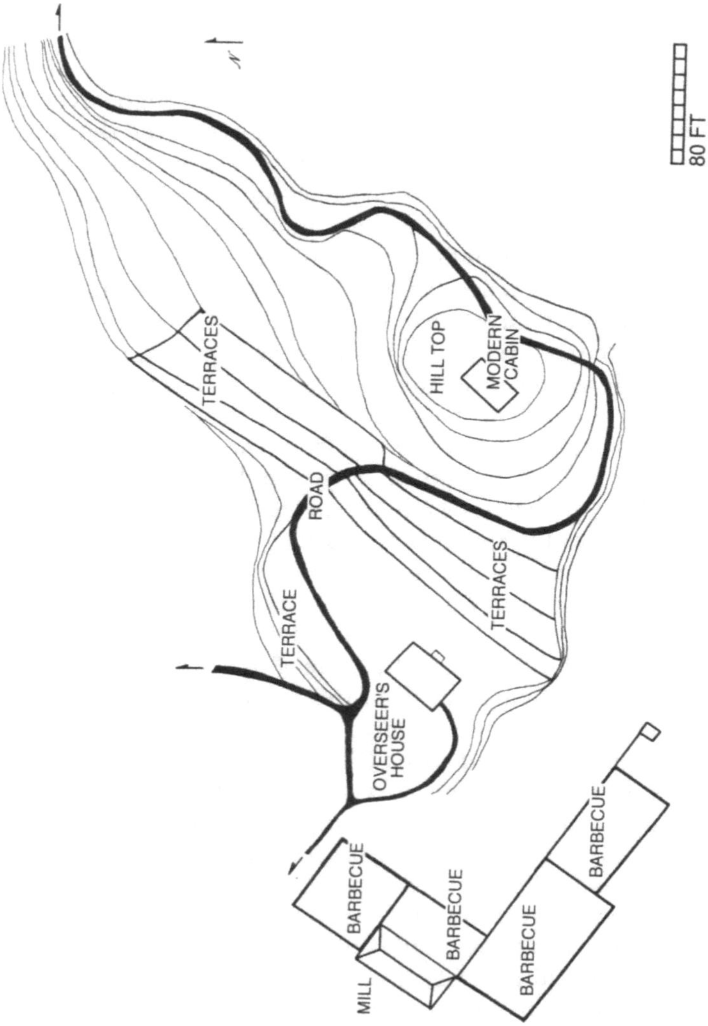

Figure 16. Clydesdale site plan.

The pulping mill at Clydesdale is located within a massive two-story stone structure, which measures 54'10" by 26'8" (see Figures 17 and 18). The ground floor of the structure contains the mill machinery; the pulping mill and waterwheel are located on the western side of the large single room composing the ground floor. The vertical overshot waterwheel was originally powered by an aqueduct that transported water from the Clyde River into the mill complex. The upper floor contains two rooms. The western room measures 20'6" by 22'10"; the eastern room measures 29'11" by 22'10". It is possible that these rooms were once occupied by members of the estate staff or one of the enslaved supervisors. Indeed, local informants call this building "the Great

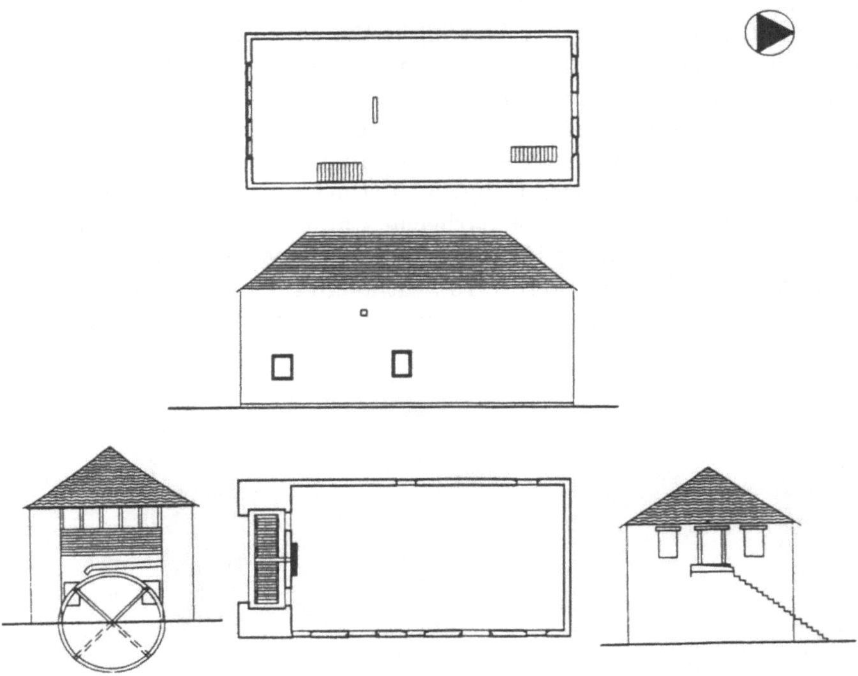

Figure 17. Architectural drawing of Clydesdale mill building

House," further corroborating the possibility that the upper story of the mill was once occupied (Figure 17).

Although the building is capped by a massively framed hipped roof, an examination of the exposed framing timbers revealed no indication that there was ever a ceiling; the upper rooms appear to have been open to the garret. Given that these rooms are located within the mill building, and that they were probably never finished, it is also possible that they were used as a warehouse space or coffee store. This usage would have been in keeping with Laborie's instructions, which suggested that the mill and the coffee store be located within the same building. This use of space conforms to the spatial organization apparent in Thomas Harrison's 1849 isometric drawing of the coffee store and mill building at Oldbury (see Higman, 1988), a plantation in the parish of Manchester. This drawing is the only known contemporary rendition of a nineteenth-century Jamaican coffee mill. At Clydesdale, the mill building and the overseer's house are built of similar construction techniques, suggesting that they are contemporaneous.

The mill is flanked to the west by a linear series of barbecues. The larger of the two measures 47' by 88'10"; the rear barbecue is somewhat smaller. Behind the smaller barbecue stands a small timber-framed pulping mill, measuring 15'10" by 17'. Water power was also used to drive this smaller mill (Figure 19). Although the interior of this building was inaccessible during field examination, the exterior cladding of

Figure 18. Clydesdale mill building.

the building exhibits circular saw markings, suggesting that this small mill was constructed at a later date than other features on the landscape, as circular saws did not come into common use in North America until the middle of the nineteenth century (Ed Hood, personal communication, 1996).

The secondary mill building may have been part of an early twentieth-century renaissance of coffee production at Clydesdale. According to a local elderly informant, sometime after the First World War a presumed British deserter named MacLeverty arrived at Clydesdale with a cache of firearms. Using these weapons, he forced the people living in and around the plantation building out of their houses and off of their land, assuming the title "Busha" (a term used for white planters following emancipation). According to this account, MacLeverty resumed coffee production at Clydesdale, expanding cultivation to the lands he had stolen from the local people. My informant reported that MacLeverty was eventually tracked down by the British authorities, who were presumably preparing to arrest him for desertion. According to the story, MacLeverty was tipped off that he was about to be arrested. Rather than surrender to the British authorities, he committed suicide in the overseer's house sometime in the 1920s. It is possible, though not yet confirmed, that this "MacLeverty" may have actually been a

Figure 19. The small mill at Clydesdale.

McClarty, perhaps a descendant of the McClartys who owned the plantation in the nineteenth century.

The mill complex at Chesterfield is in a much worse state of preservation than that at Clydesdale. Nothing remains of the pulping mill save a foundation (Figures 20 and 21). From this scanty evidence, however, several observations about the material space of the mill can be drawn. The ruins of an aqueduct indicate that water from the Fall River was diverted to power the mill and to fill the several cisterns surrounding the foundation ruin. As was the case with Clydesdale, the mill was a distinct structure, located away from the overseer's house. A series of barbecues is located behind the mill complex. Unlike the Clydesdale barbecues, which were located on level ground abutting the Clyde River, the Chesterfield barbecues are terraced into a hillside that overlooks the confluence of the Yallahs River and the Fall River.

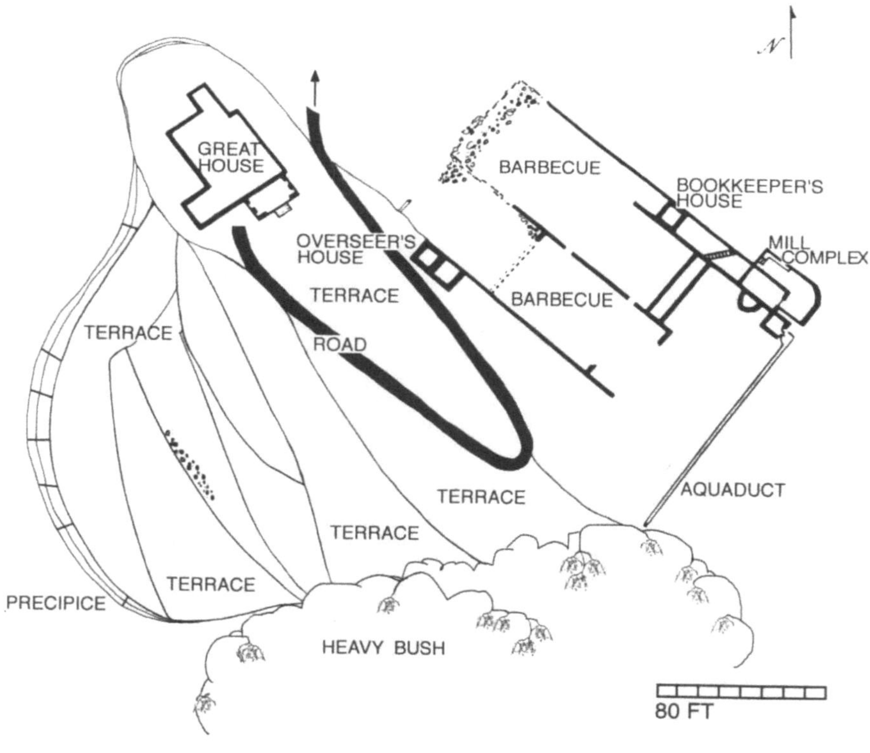

Figure 20. Chesterfield site plan.

The original pulping mill at Sherwood Forest differs from the other two discussed above, as it was actually located in the overseer's house (Figure 22). The mill was housed in one room of this structure, while the overseer occupied the other. Like the mills at Clydesdale and Chesterfield, the pulping mill was water powered. However, unlike the other two estates, Sherwood Forest is not situated *below* a river, but rather is located *above* the Negro River. To compensate for the lack of river-fed water power, the mill complex is attached to a spring-fed cistern, which serves as a sort of "mill pond." During the pulping process, water from the cistern was channeled to the mill by means of a small aqueduct, which survives today, but is not functional. The cistern itself now serves as an occasional swimming pool for the Deichman family, the current owners of the estate.

The barbecues at Sherwood Forest are similar to those at Chesterfield, as they are terraced into a hillside (Figure 23). Above the terraced barbecues is a series of smaller terraces, which are currently used as a coffee nursery. It is possible that this is the original purpose of these terraces, as coffee plants require careful nurturing for several

Figure 21. Ruins of the mill complex at Chesterfield.

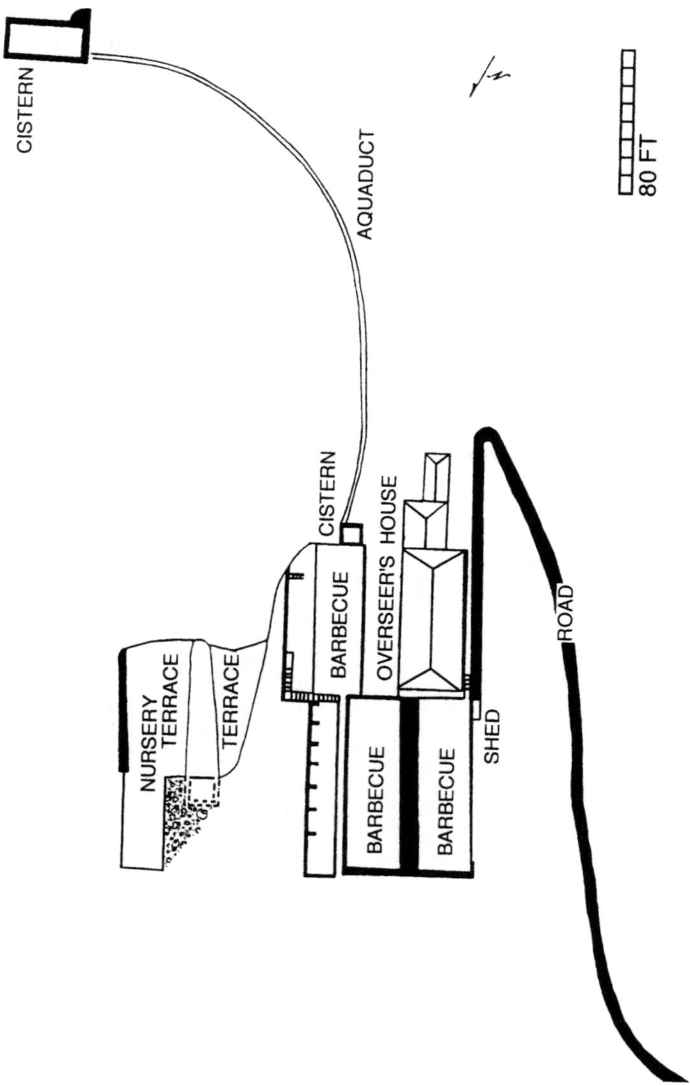

Figure 22. Sherwood Forest site plan.

Figure 23. The barbecues at Sherwood Forest.

years before they can be transplanted into the coffee fields. It is likely that each plantation had some kind of nursery facility.

The Material Space of the Elites

Although the European population on each coffee plantation would be small, there was nevertheless a distinct class structure in place incorporating the planter-owners, the overseers, and the book-keepers. The class structure was reified by the spaces reserved for members of these strata; this class-stratified space was manifested most directly in the creation of the material spaces of overseers' houses and great houses.

As was the case during the actual production of coffee, the production of plantation space was supervised by a white employee of the estate, usually the overseer. During slavery, the overseer was often the de facto manager of the estate, responsible for organizing labor and supervising the production and distribution of coffee as an agricultural commodity (Laborie, 1798). As such, the overseer played a crucial role in the work life of slaves. It was the overseer who would determine what work was to be done by whom, and it was he who would either supervise or directly mete out punishment on the enslaved laborers (Stewart, 1971 [1808]:129–130; Walvin, 1994:239; for a particularly graphic description of punishment meted out by an overseer, see Hall, 1989:72–75).

As supervising managers on the estates, overseers were often granted quarters of their own. Overseers served as intermediaries between the truly elite planter-owners and the primary producers; their houses both reflected and created this intermediate class relationship. To interpret the material space of this class of supervisor, we can turn to the archaeological record of the Yallahs region, which retains several examples of overseers' houses in various stages of preservation.

Perhaps the best example of an extant overseer's house is that located on Clydesdale (Figures 24 and 25). The Clydesdale overseer's house, which is currently called "the Rest House," is located within the industrial complex of the plantation. This is a two-story building. There is no interior stairwell connecting the two stories, however, suggesting that the spaces of the two stories were used for separate purposes; the only access to the second story is an external staircase, originally constructed of wood. The upper story is finished, suggesting that this area was the domestic space of the overseer, while the lower story is not, suggesting that this area was used either for the storage of tools and equipment, or for the storage of coffee.

The upper story is divided into three rooms with a second-floor veranda, now boarded over. It is likely that the largest of the three rooms served as a social area in which the overseer would entertain guests and possibly those slaves who had business to discuss with him. One of the smaller rooms was probably a bedroom for the overseer and his wife (if he had one). The second bedroom may have been used by their children, or perhaps another member of the white estate staff. The veranda overlooks the coffee works, allowing the overseer the ability to supervise the pulping and drying of coffee without having to leave his house. The walls of this structure are cut stone, and are 2 feet thick. Clearly, the overseer had the ability to shut himself and his family into this structure with a significant buffer between them, the elements, and the enslaved workers. With the exterior stairway the only access to the domestic quarters of the upper story, the overseer lived in a virtually impenetrable fortress.

The overseer's house at Sherwood Forest also survives today. This particular structure exhibits two phases of construction, with the older house actually serving as part of the foundation for the later house. As was the case with Clydesdale, the earlier overseer's house is a relatively small structure built with massive stone walls. This structure is divided into two rooms; unlike the overseer's house at Clydesdale, the one at Sherwood Forest contained the domestic space of the overseer as well as the pulping mill. In this case, the overseer would be in direct contact with the first stage of the industrial processing of coffee. The other key architectural feature of this building is a veranda, oriented

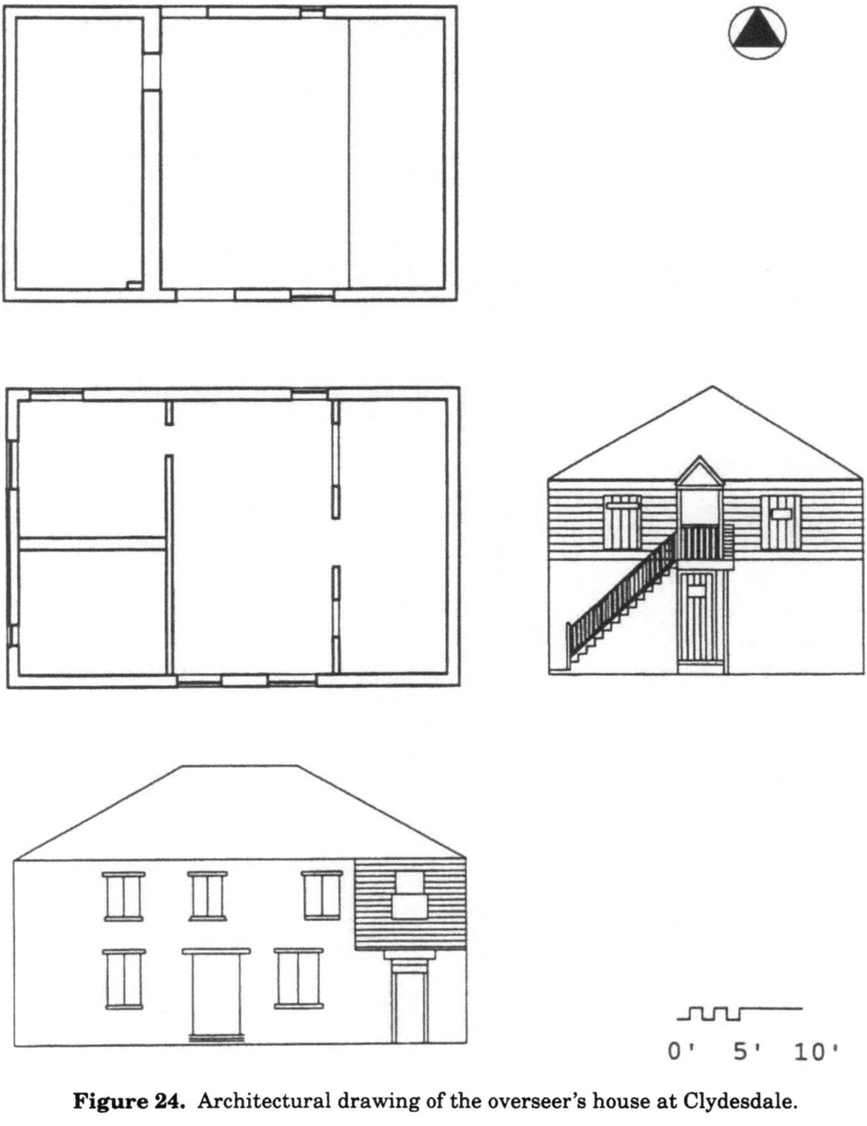

Figure 24. Architectural drawing of the overseer's house at Clydesdale.

Figure 25. The Clydesdale overseer's house.

toward the barbecues. From this vantage point, the overseer would have the capability to be in direct surveillance of the drying beans. At Sherwood Forest, the domestic space of the overseer, at least during the first phase of the plantation's existence, was inseparable from the space of production (Figures 26–28).

Yet another overseer's house survives at Chesterfield. As mentioned above, the preemancipation Yallahs coffee complex experienced two separate phases of development, the first coming in the decades surrounding the turn of the century, the second some 20 years later. At Chesterfield, the small building identified as the overseer's house exhibits two distinct phases of construction, coinciding somewhat with the two phases of coffee production. In its first phase, this was a squat, two-story building, with one room on each floor. The interior rooms each measure 10' square; the walls are made from fieldstone covered with lime plaster (Figure 29).

Both this building and the plantation experienced a second phase of construction, probably beginning around 1824. The crop accounts for that year indicate that the plantation management was hiring a number of skilled craftsmen from neighboring plantations. Approximately £140 were spent to hire the labor of carpenters and masons from Penlyne Castle, Pleasant Hill, Mocho, Petersfield, and Sherwood Forest, all neighboring coffee plantations. In 1825, an additional £30 were

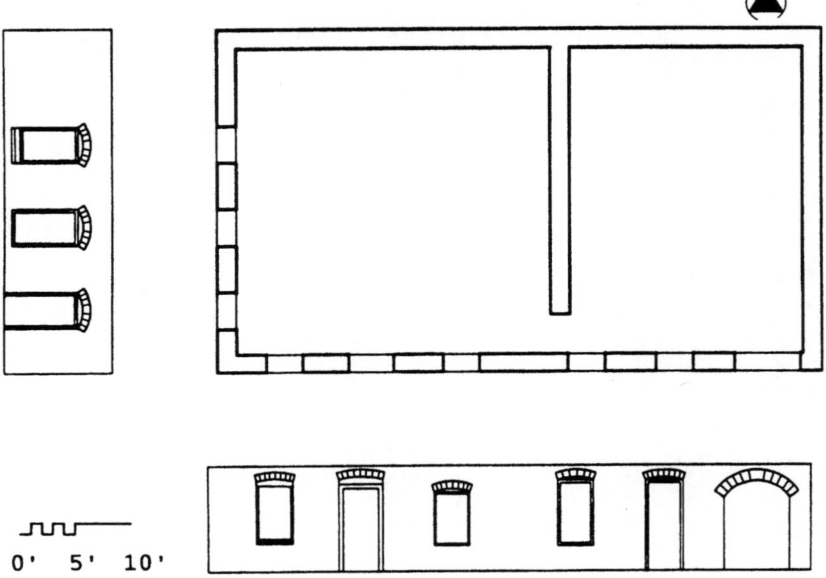

Figure 26. Architectural drawing of the mill/overseers house at Sherwood Forest.

Figure 27. The exterior of the mill/overseers house at Sherwood Forest.

Figure 28. The interior of the mill/overseers house at Sherwood Forest.

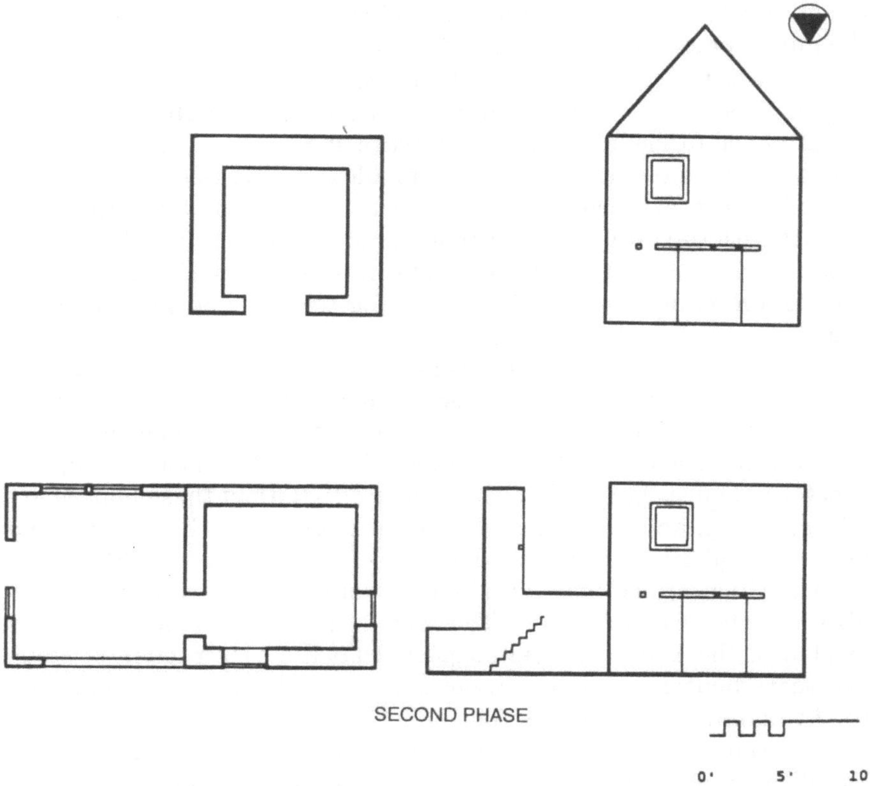

Figure 29. Architectural drawing of the Chesterfield overseer's house.

spent to hire artisan labor; the plantation also purchased 10 hogsheads of lime, most probably for construction.

Probably as part of this construction, the now heavily overgrown overseer's house was expanded. An additional 10' by 12' room was added to the second story, atop a massive stone foundation. The walls of this new room were far more gracile than the massive fieldstone walls of the original, being a mere 7" thick and constructed with a technique that local informants described as "nag" — which is little more than a sophisticated form of wattle and daub. The wattle is a mesh of iron wire, which would be attached to studs. A type of lime mortar or concrete would be poured between the studs, forming, in effect, a 7" plaster wall. The remains of the Chesterfield great house, located nearby, demonstrate that this building was also constructed with the nag technique. It seems likely that given the expense of hired labor in the mid-1820s, the two buildings indicate a considerable investment in the improvement of elite space at Chesterfield.

As was the case at Sherwood Forest, the overseer at Chesterfield was in direct contact with production. Although the pulping mill was not attached to the house as was the case at Sherwood, the overseer's house was oriented in such a way as to allow the overseer to supervise the coffee crop as it dried on the barbecues. He would also be in visual contact with the mill, so that although not as intimately linked to the production process, the overseer was indeed in a position to watch production as he saw fit, without having to leave his own domestic space.

Although it is possible that these structures were massively built out of stone to protect the overseer's from hurricanes, an ecological-based interpretation for the massive construction falls somewhat short. At least two nearby plantations, Abbey Green and Whitfield Hall, had overseers' houses built of timber (Figure 30). Both of these houses are still standing, and are actually in better shape than the Chesterfield house. In addition, the second phase of construction at Sherwood Forest, which will be dealt with in more detail in the next chapter, was also a timber-framed house. The fact that timber-framed buildings existed in the Yallahs region suggests that the construction of stone overseers' houses was a choice, not a necessity, based, for example, on the availability of raw materials.

In contrast to the number of standing overseers' houses in the Yallahs region, there are very few surviving great houses. This may be a reflection of the relative isolation of the great houses from the rest of the plantation. For example, the locations of the great houses for both Sherwood Forest and Whitfield Hall were pointed out to me by local residents; neither is still standing, and neither site is in direct view of

Figure 30. Front elevation of the overseer's house at Whitfield Hall.

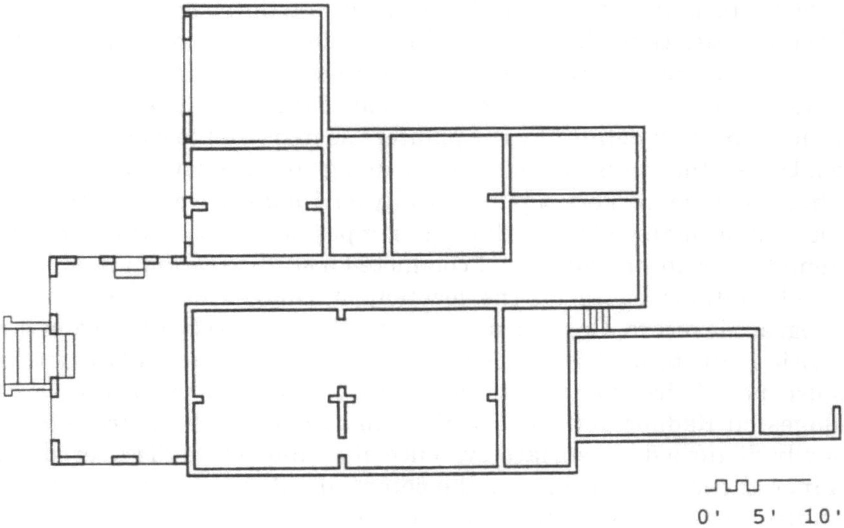

Figure 31. Floor plan of the great house at Chesterfield.

the coffee works. Of the plantations examined for this study, only Chesterfield had visible ruins of a great house.

The great house at Chesterfield was a much larger structure than any of the overseers' houses, reflecting the class differences between the estate owners and the white estate employees (see Figure 31). The house was located at the top of a ridge, in view of the Yallahs River, the overseer's house, and the coffee works. The floor plan of the great house indicates that the structure contained an 18' by 15'6" hall, three chambers, and an attached kitchen with a storeroom. A small 9' by 4' room with its own attached veranda may have served as an office. Two additional rooms in the back of the house, near the kitchen, may have served as the quarters for domestic servants. This house also had a veranda, which overlooked the river valley, the industrial works, and the coffee fields which the cartographic record of Chesterfield indicates were located on the opposite side of the river.

The Material Space of the Enslaved

As discussed in earlier chapters, plantation spaces were segregated into distinct areas. Heuristically, the material spaces controlled by the enslaved laborers can be divided into domestic space, i.e., the space in which the African population lived, and the space of production of commodities controlled by the Africans, i.e., provision grounds.

As was the case with sugar plantations, the African workers in the Yallahs region tended to live in nucleated villages clustered in areas on the plantation defined by the planters as marginal to coffee production. Unlike the industrial and elite housing on the estates, the domestic houses of the workers, again typically for Jamaica, were constructed of perishable materials, and thus were impermanent structures. In the attempt to locate the villages, I conducted a surface reconnaissance of five plantations based on the location of villages indicated by the cartographic record. Comparing location information obtained from the historic maps to modern topographic maps and the actually existing landscapes, I first located the areas that I believed were the slave villages on Radnor and Clydesdale. Unfortunately, these areas were heavily disturbed; no surface evidence of a village was discovered. At Radnor, the areas identified as the potential village sites were planted in coffee; again, no evidence of nineteenth-century occupation could be discovered without disturbing valuable modern coffee trees. The cartographic records for Whitfield Hall, Sherwood Forest, and Chesterfield do not indicate the location of slave villages. Archaeological reconnaissance again proved futile, as heavy coffee production at Whitfield Hall, anthurium beds at Sherwood Forest, and the division of Chesterfield into a housing project, what is colloquially known as a government housing "scheme," limited the areas available for reconnaissance. No evidence of occupation was located. The attempt to discover the locations of the slave villages were thwarted not only by the relative silence of the surface record, but also by heavy soil erosion. These troubles were further compounded by the fact that with the exception of Chesterfield, all of the Yallahs plantations considered in this study are once again in cultivation. The Blue Mountain region is experiencing a modern coffee boom; nearly every conceivable corner of the old plantations is once again under coffee cultivation, including the areas I believe were once villages. I was thus unable to conduct any subsurface testing without disturbing coffee trees.

The sketchy representations of slave villages in the cartographic record further complicate these methodological limitations. Few of the preemancipation maps from the Yallahs drainage depict slave villages. More commonly, a parcel of land within the plantation is referred to as "Negro grounds," suggesting that the surveyors and their patrons, the planters, took less interest in the spatial organization of the slave villages than, for example, the coffee fields. This absence may represent an accepted division of space within the plantation. The maps suggest that at least to some degree, the planters did not attempt to exert specific spatial control over the slave villages; the slaves themselves were in possession of these lands and houses. This situation changed radically following emancipation, when a far greater proportion of the maps that survive do indeed show the location and layout of settlements. This reflects a changing trend toward a more severe control over space following emancipation, a trend that will be discussed in more detail in the next chapter.

Despite these limitations, there are some conclusions that can be drawn about the organization of slave villages on coffee plantations in the Blue Mountains. In an analysis of the cartographic record of Jamaican coffee plantations, Barry Higman offers a few conclusions relevant to this study, though it should be noted that his study included maps from the entire island, spanning the dates 1780–1860. Higman's study was primarily a statistical analysis designed to produce generalizations about the internal structure of Jamaican coffee plantations. He did not provide a list of the 61 maps used in calculating his statistics; thus, with the single exception of Radnor, which he discusses in some detail, it is impossible to determine if any Yallahs estates were included in his study. In analyzing his sample, Higman calculated that the mean size of a slave village on a Jamaican coffee plantation during this 80-year period was 2.7 acres; he asserted that this relatively small size did not change much through time (Higman, 1986:81). Higman suggested that the mean distance between the works and the settlements decreased over time. In his analysis he provided mean distances between works and villages per decade, which are reproduced in Table 1. As he cautioned in this article, however, the sample size for plans that actually depict both works and villages for any given decade is small. Unfortunately, he did not offer an analysis of the internal structure of the villages.

Table 1. Mean Distance between Works and Villages on Coffee Plantations, Calculated by Higman (1986)

Decade	Mean distance (yards)	Standard deviation
1790–99	366	172
1800–09	326	410
1810–19	299	200
1820–29	178	99
1830–39	231	89
1840–49	209	141
1850–59	139	71

Very few archaeological studies have looked specifically at the internal arrangement of slave villages within Jamaican plantations. The most significant study was that conducted by Douglas Armstrong on Seville Estate, a sugar plantation located on the littoral of the north coast, in the parish of St. Ann. Armstrong was able to locate two distinct slave villages, one dating to the early eighteenth century, the other to the late eighteenth century. Using cartographic information supplemented by excavation of numerous slave houses, Armstrong has been able to discuss how the spatial arrangement of the villages differed. The earlier of the two villages was organized linearly. It was composed of four symmetrically organized rows of houses, two located on either side of a road. The later village consisted of a number of houses organized around a common, reflecting a more dispersed circular village. Armstrong concludes that the settlement pattern of the later village reflects a situation of decreased surveillance of the slaves. Noting that the great house experienced a phase of reconstruction at approximately the same time as the later village was constructed, Armstrong has concluded that the planter elites may have been too preoccupied with the reconstruction of the house to dictate the construction of a symmetrically organized village (Armstrong, personal communication, 1996; Armstrong and Kelly, 1990).

Using Armstrong's conclusion as a basis for deduction, one can hypothesize that the settlement pattern of villages on coffee plantations should more resemble the later slave village at Seville than the earlier. It has been suggested that the daily labor regime of working coffee plantations was less severe than on sugar plantations (e.g., Thome and Kimball, 1838:406–407). Yallahs coffee plantations were constructed over a relatively brief period, during which time significant forest clearing, terracing, planting and construction of works and elite housing would have been simultaneously accomplished. Furthermore, many of the coffee plantations in eastern Jamaica were carved on steeply sloped hillsides; some 50% of coffee plantations were located in areas with mean slopes exceeding 20% (Higman, 1986:77). Thus, one can hypothesize that because the investment of capital and attention into the construction of productive space was high, the coffee planters would have left the construction of villages to the slaves. Furthermore, the rugged terrain undoubtedly inhibited the construction of symmetrically ordered villages.

Unfortunately, this hypothesis is difficult to either prove or disprove. No slave villages have yet been identified in the Blue Mountains. The cartographic record of slave villages is poor, for as Higman notes, the purpose of rendering plantations into plans was part of a strategy to control and monitor the production of coffee. The layout of slave villages was a secondary consideration to the quantification of production (Higman, 1986:76).

Nevertheless, additional observations can be made about the layout of slave villages based on an analysis of two cartographic representations of preemancipation villages. The first is for Radnor, which is based on the redrawn plan of the estate published by Higman (1986, 1988). Recall that Higman calculated that this village occupied 2.7 acres of meadow. On his representation of Radnor, Higman has rendered the village as a cluster of 30 buildings located in a meadow (Higman, 1986). According to the St. David vestry accounts, in 1806 the Radnor slave population totaled 252. If the plan accurately depicted the number of houses, if not their specific location, this would suggest that approximately 8 people were living in each of the houses. Regardless of the accuracy of the number of houses, if all 252 people were indeed living in the 2.7-acre meadow, the slave village would have been a very densely populated place indeed, with a population density of approximately 93.33 people per acre. It should be noted that this densely populated zone only incorporates the area encompassing the slave village. It is likely that many people spent their off-duty hours in the provision grounds, which will be discussed below.

The only other representation of a slave settlement on a map of a Yallahs region coffee plantation that can be confidently dated to the preemancipation period is STA 920, a depiction of Clydesdale plantation, dated 1810. In this representation, a total of 11 houses are depicted; 7 of these are located along one side of a road. Three others appear clustered some distance away. According to the legend on this map, these houses were located in a 10-acre coffee field. In 1818, the slave returns for Port Royal indicate that 109 slaves were attached to the plantation at that time. If we accept that there were ten slave houses on the estate, this information suggests that there were approximately 11 slaves per house located on the estate.

Until further archaeological research identifies the location of a slave village in the Yallahs region, only preliminary observations about slave villages can be made. The villages consisted of a number of houses clustered together relatively close to the coffee works. There were relatively few houses depicted per person, suggesting either that the houses were very crowded when inhabited or that some of the people did not spend much time in the houses. As a caveat, Higman suggests that the high ratio of people to house depicted on Jamaican plans at large may indicate that the surveyors did not accurately depict the number of houses located on the estate (Higman, 1988:244).

It is most likely that slave houses in the Blue Mountains were constructed of wattle and daub with thatched roofs. Entries for the first few weeks of the Radnor plantation book indicate that while three or four people were "making huts," one person was cutting thatch, suggesting that the thatch was used during house construction. An undated lithograph depicting a "mountain cottage scene" in Jamaica corroborates the suggestion that houses located in the mountains were constructed with thatched roofs (Figure 32). This illustration suggests that at least some houses contained two rooms, here indicated by a secondary roof covering an "el" attachment. Matthew Lewis described the slave houses on his sugar estate as being constructed of wattle and daub (Lewis, 1929 [1834]:197). It seems likely that the slave houses in the Yallahs region conformed to these standards.

Provision grounds were arguably the most important space to the enslaved populations. The plots that were gardened not only provided food for the sustenance of the workers and their families, but were spaces in which the enslaved were free to produce surplus for sale. As Mintz and Hall (1960) have argued, preemancipation Jamaican planters with estates located in the interior or in mountainous areas relied on the production rather than the importation of food to feed their laborers. The laborers, in turn, were allowed access to lands marginal to export commodity production, to produce their own foodstuffs, and significantly, were allowed to exchange any surplus in markets.

Mintz and Hall (1960) have suggested that each slave, or more accurately, the recognized head of a slave household, was allotted a small plot in the provision grounds, which on sugar estates were sometimes located miles from the slave villages. This argument is supported by observations made by William Beckford in 1790 and J. Stewart in 1823. Beckford observed that a "quarter acre . . . will be fully sufficient for the supply of a moderate family, and may enable the [slave] to carry some to market besides" (Beckford, 1790:257). Stewart elaborates on this: "Each slave has . . . a piece of ground (about half an acre) allotted to him as a provision ground. This is the principal means of his support; and so productive is the soil, where it is good and the seasons regular, that this spot will not only furnish him with sufficient food for his own consumption, but an oversurplus to carry to market. . . . If he has a family, an additional proportion of ground is allowed him, and all his children from five years upward assist him in his labours" (Stewart, 1969 [1823]:267).

Although an extremely important space to the laborers, specific provision grounds are perhaps the most difficult material space to identify and analyze archaeologically. As they tended to be located away

Figure 32. A nineteenth-century lithograph of a mountain cottage scene in Jamaica. Courtesy of the National Library of Jamaica.

from the centers of the plantations, and may not have had any permanent built landscape features associated with them, it is all but impossible to find these spaces today.

As the provision grounds were a space occupied and utilized by the slaves for their own use, little information on preemancipation provision grounds appears in the cartographic record of the Yallahs region. On Pechon's 1808 rendition of Mavis Bank, no subdivision of the "Negro Grounds" is rendered, suggesting that the surveyor either was not interested in the exact allocation of land, or else did not gain access to this particular area of the plantation. It is also possible that given the nature of the occupational use of the space, the planters were not interested in the exact allocation of the land, but left the allocation and occupation of the land to the slaves themselves. Of the estate's 302 acres depicted on the map, 79¾ acres, approximately 26% of the entire acreage, were identified as "Negro Grounds and Provisions."

Pechon's depiction of Radnor plantation, reproduced by Higman (1986, 1988), gives some further indication of the position of provision grounds in the division of material space on coffee plantations. Of the estate's 689 acres, 133 acres were in Negro grounds and provisions, and an additional 222 acres were in woodland and Negro grounds. According to Higman's rendition of the plan, there were four separate areas of the plantation designated as Negro grounds, and a fifth identified as woodland. As was the case on Mavis Bank, most of this land was located on the material periphery of the estate. On Radnor, the distance between the slave village and the Negro grounds and provision grounds was between one-third and three-quarters of a mile.

The spatialities of these new landscapes defined how these spaces were experienced both by the enslaved Africans and by the European planters. The next section discusses how these spaces were experienced, how they were intended to form social relations of production, and how such manipulations were resisted.

SPATIALITY OF ENSLAVED LABOR AT RADNOR

The shells of the material spaces that comprise the archaeological record of the coffee plantations tell us little about the social lives of the people who lived and worked there. In order to interpret how people interacted in these spaces, we must turn to historical documents. The best source for the interpretation of social spaces for the Yallahs region is the Radnor plantation book. A close interpretation of this manuscript volume sheds light on how these spaces were lived and experienced. In this section, I will use this data source to analyze how the enslaved laborers experienced the various spaces of the plantation.

Spatialities of the Coffee Fields

The Radnor plantation book documents the life of the plantation from January 1822 through February 1826. As described in Chapter 4, there was a distinct division of labor for the enslaved population on coffee estates. The role that an individual played in this division of labor determined which spaces she or he inhabited, what activities were pursued in the spaces, and the relationships experienced with other members of the community, both on and off the plantation. The division of labor on the first workday of each year during this period is reproduced in Table 2. Those working in the gangs, the field laborers, would spend their workdays in the coffee fields, hoeing, picking coffee, pruning trees, weeding, or accomplishing whatever task was required in the fields on a given day. As was true with most plantation workers, the field hands labored for 11 days per fortnight. Sundays and alternate Saturdays were considered "free" days for the slaves; according to the plantation book, the people would spend this time working in their provision grounds.

Men and women would work side by side in the coffee fields, under the eye of the drivers. The composition of the gangs would differ from month to month and year to year as older people joined the second or third gang and younger people were moved to the first gang. The plantation book contains a list of the slaves on the plantation as of January 1825, which includes their occupation. This list allows us to interpret the gendered composition of the gangs at that time. The first gang was supervised by three slave drivers: Monday, who is identified as the "Head Driver," and Hector and Cato, who are identified as the "Second Driver" and the "Third Driver," respectively. When the list was drawn up, Monday's gang contained 30 men and 52 women, suggesting that the bulk of the heaviest plantation work was done by women. The second gang, supervised by Seipio, contained 13 men and 9 women, although one of the women, Rachael, is identified as a cook. The third gang was driven by Sampson and contained 12 males and 9 females; Princess is identified as the cook for this gang.

Spatiality of Provision Grounds

As the purpose of the Radnor plantation book was to monitor and record the production of the coffee plantation, it is relatively silent

Table 2. Number of Workers per Occupation at Radnor Plantation, at the Beginning of Each Year, 1822–1826

Occupation	Jan. 1, 1822	Jan. 1, 1823	Jan. 1, 1824	Jan. 1, 1825	Jan. 2, 1826
1st gang	49	68	56	63	67
2nd gang	22	25	20	16	17
3rd gang	28	14	21	19	18
4th gang	16	0	0	0	0
Carpenter	4	3	5	5	5
Mason	3	3	3	3	3
About works	4	4	3	3	4
In hospital	12	14	14	6	7
Children	22	24	41	41	40
Various	2	2	9	3	5
In town	3	2	1	1	1
Invalids	8	8	6	7	7
Doctors	2	2	1	1	1
Domestics	7	7	6	6	6
Pregnant	0	4	3	2	2
Confined women	0	0	0	2	3
Midwife	1	1	1	1	1
Nurses	1	3	3	3	2
Runaways	1	1	1	1	2
Day off	5	3	3	4	0
Working stock	1	2	1	1	1
Minding stock	3	3	4	4	4
Got yaws	19	19	7	5	9
Sawyers	3	3	3	5	3
At great house	1	1	1	1	1
Watchman	5	6	6	7	7
Kings Road	0	0	0	2	0
Absent	0	0	0	0	0
Making huts	3	0	0	0	0
Cutting thatch	0	0	0	0	0
Total	226	223	220	213	218

Source: Radnor plantation book.

about how the enslaved African population experienced the spaces of the provision grounds. The journal does faithfully record that on Sundays and alternate Saturdays, the enslaved were either "cultivating their provision grounds" or else were "employed in their provision grounds." What is noticeable about the majority of these entries is that the bookkeepers used the possessive adjective "their" when describing the provision grounds. This indicates that the provision grounds were recognized as a dual space. While the actual land was owned by the plantation, the spaces within the provision grounds were recognized as belonging to the enslaved workers. This interpretation is reinforced by several other entries in the plantation book concerning the commodities produced by the laborers on their provision grounds and their yards. For example, in the plantation accounts, there are numerous entries concerning the purchase of provisions for the estate staff from the enslaved workers. In the account for John Cherrington, the estate attorney or manager, the most common commodities purchased from the workers were fresh pork and castor oil. Table 3 lists all of the commodities purchased by the estate from known individuals. As is seen, several of the workers were involved in a commodities trade with the estate, particularly for food stuffs, especially pork, cocoas — a squash-like vegetable — and castor oil. There are numerous other entries in the estate book indicating that the white staff purchased other foodstuffs, probably from the enslaved workers, without information as to the individual from whom purchased.

Although the evidence is scanty, it seems likely that at least some of the working population spent more time producing commodities for sale than others. For example, Thomas Kelley, who is identified in the plantation slave list as a doctor, sold £7.18.7 worth of commodities to the estate between January 1822 and October 1824. Although this was not a vast amount of money, it represented more than one month's wages for the lowest paid member of the white estate staff, the bookkeeper, who the plantation book indicates was paid £60 per annum, or £5 per month. Thomas Kelley appears more often in the accounts than any other individual. He provided pork and castor oil to the estate. Monday, the head driver, sold cocoas to the estate; unfortunately, the ledger does not indicate how much these transactions were worth. Pork was sold to the estate almost exclusively by men; of the 584.25 pounds of pork purchased by the estate from known individuals, all but 19 pounds were bought from men. Of the women on the estate, only Rebecca sold pork, and that was on only one recorded occasion. Castor oil was a commodity that was far more frequently purchased from women; in all, nine separate women made and sold castor oil to the estate during the years covered by the plantation book.

Provision grounds remain the most difficult spaces to interpret. These were areas recognized during the days of slavery as belonging, in some sense, to the enslaved Africans who worked on them. It was in these spaces that men and women toiled for themselves and their families, to provide food for sustenance, and surplus for exchange, both to the estates and in the internal market economy. Unfortunately, the Radnor plantation book records very little about the internal markets of the Yallahs region. We do not know, beyond conjecture, if any of the

Table 3. Commodities Purchased from Known Individuals at Radnor, 1822–1826

Date	Purchased from	Commodity	£	s	d
1/17/1822	Thomas Kelley	30 lb pork	1	5	0
2/6/1822	Nanny	1 bottle oil		3	4
2/6/1822	Pompy	2 bottles oil		6	8
2/23/1822	Dick	11½ lb pork		9	7
3/16/1822	Thomas Kelley	3 bottles oil		10	0
4/6/1822	Duncomb	37½ lb pork	1	11	3
4/19/1822	George	28½ lb pork		16	3
5/4/1822	Robert	16 lb pork		13	4
5/6/1822	George	26½ lb pork	1	2	1
6/21/1822	Phillis	2 bottles oil		6	8
7/15/1822	Davey	45 lb pork	1	17	11
8/21/1822	Nanny	2 bottles oil		6	8
9/17/1822	Robert	39 lb pork	1	11	8
9/29/1822	Robert	26 lb pork	1	0	5
10/1/1822	Thomas Kelley	2 bottles castor oil		6	8
10/10/1822	Thomas Kelley	31 lbs pork	1	2	1
11/6/1822	Thomas Kelley	2 bottles castor oil		6	8
11/13/1822	Thomas Kelley	23¾ lb pork		10	2
2/9/1823	Monday	3 bushels cocoas			nd
3/22/1823	Rodney	31½ lb pork	1	6	3
4/18/1823	Nanny	2 bottles castor oil		6	8
4/13/1823	Cornwallis	19 lb pork		15	10
4/26/1823	Thomas Kelley	58 lb pork	2	18	4
5/12/1823	Monday	1 basket cocoas			nd
5/1823	Monday	5 baskets cocoas			nd
6/7/1823	Mary	2 bottles castor oil		6	8
6/1823	Monday	3 bushels cocoas			nd
7/1823	Monday	2 bushels cocoas			nd
7/20/1823	Old George	26 lb pork	1	1	8
9/5/1823	Quomin	14 lb pork			nd
12/27/1823	Nanny	1 bottle castor oil		3	4
6/20/1824	Moriah	2 bottles castor oil		6	8
7/5/1824	Thomas Kelley	2 bottles castor oil		6	8
8/11/1824	Penelope	2 bottles castor oil		6	8
9/6/1824	Nanny	2 bottles castor oil		6	8
10/2/1824	Thomas Kelley	28½ lb pork	1	3	0
10/16/1824	Princess	2 bottles castor oil		6	8
10/24/1824	"the Negroes"	fowls	1	8	4
11/23/1824	Penelope	2 bottles castor oil		6	8
12/20/1824	Rainsford	32 lb pork	1	6	8
2/2/1825	Pinkey	4 bottles castor oil		13	4
2/10/1825	Affu	8 lb pork		6	8
3/28/1825	Penelope	2 bottles castor oil		6	8
4/29/1825	Robert	9 lb pork		7	6
5/9/1825	Christianna	2 bottles castor oil		6	8
6/13/1825	Rachael	2 bottles castor oil		6	8

Table 3. (*Continued*)

Date	Purchased from	Commodity	£	s	d
7/18/1825	Christianna and Mary	2 bottles castor oil		6	8
8/9/1825	Penelope	2 bottles castor oil		6	8
8/13/1825	Rainsford	24½ lb pork	1	0	5
8/22/1825	Penelope	2 bottles castor oil		6	8
9/12/1825	Roseline	2 bottles castor oil		6	8
10/26/1825	Mary	2 bottles castor oil		6	8
11/21/1825	Penelope	2 bottles castor oil		6	8
1/2/1826	Doncan	1 bottle castor oil		3	4
2/19/1826	Rebecca	19 lb pork			nd

Source: Radnor plantation book.

people in the remote interior were able to participate in the market system. Radnor, like many of the Yallahs estates, was located in a remote area far from any market town. While it is possible that the enslaved workers did travel to market to sell the commodities they produced on the provision grounds, no record of such activity in the preemancipation Yallahs period has yet been uncovered.

INTERSECTIONS OF SPACE: THE SPATIALITIES OF CONTROL AND RESISTANCE

The interpretations of cognitive space and the actions of social space within material space together compose spatiality. As I argued in Chapter 2, members of different social groups, particularly in a stratified society, will experience spatialities differently. In the context of the Yallahs region coffee plantations, spatialities were manipulated by the planter class as part of a larger strategy of social control over the lives and labor of the African Jamaican population. Under the slave regime, these exertions of power and control were extreme, and often brutal. However, as Hodder (1985) via Gramsci has argued, no exertion of control over people's actions can be absolute. Just as the spatialities of control were designed to oppress African Jamaicans forced to toil on coffee estates, so too did spatialities of resistance develop in the struggle against this oppression.

The Spatialities of Control

Numerous variations on the spatialities of control were developed in preemancipation Jamaica, and were utilized in the Yallahs region.

Among those that can be interpreted from the intersections of the historic and archaeological records are the attempts to control movement through space, which I call the spatialities of movement, and the attempts to control action through panoptic Benthamite surveillance techniques, which I call the spatialities of surveillance (Delle et al., in press).

One of the methods by which Europeans attempted to control the African Jamaican population was by restricting free movement through space by the enslaved population — the spatiality of movement. To examine this aspect of the spatiality of enslaved labor, we can analyze the freedoms and restrictions of movement experienced by the working population. For this analysis, I will again turn to one of the most complete primary documentary sources for the Yallahs region, the Radnor plantation book.

The interpretation of the spatiality of movement from an estate book, like any similar historical exercise, is by default biased from the perspective of those in power, who, after all, recorded the history. From the perspective of the white estate staff, therefore, recorded movement through space can be interpreted as having occurred in two categories: sanctioned movement and illicit movement. Certain members of the plantation work force had greater latitude to move through space than others, by virtue of the role they played in the plantation work force. For example, groups of workers were expected to transport the processed coffee crop to the wharves for shipment. The men entrusted with the movement of the coffee from the estate to the waterfront were allowed a sanctioned movement through space. Absconding from the plantation, or running away, was a spatiality of movement that was not sanctioned, and will be discussed in more detail below.

Several of the members of the Radnor community were allowed sanctioned movement through space. For example, on January 21, 1822, Dunkin, a field worker from the first gang, was reported to be "sick in town, and not returned with the wharf book and bags." Several days later, on the 24th, "Dunkin returned . . . from town." By town, as is still the case in Jamaican vernacular, the estate book was referring to Kingston. Dunkin was among those on the plantation who were allowed some free movement between Kingston and Radnor. His movement was sanctioned by the estate, as it was related to business, in this case, keeping accounts with the wharves from which Radnor coffee was shipped.

Similarly experiencing sanctioned movement through space, small groups of Radnor workers traveled on business regularly to the outskirts of Kingston. The plantation journal kept tight records on how many people went on these trips, how many pack animals were brought with them, when they left and when they returned. During each month from March to October 1822, between one and three overnight trips were taken by groups of 6 to 12 people to Hope estate, near Kingston, with coffee; the comings and goings of these groups were closely recorded. In 1823, the traveling season was somewhat shorter, ending in August rather than October. In April of that year, the Radnor managers began to record the names of the people traveling with the coffee to Hope. At the time, Hope was an estate at the foot of the Blue Mountains on the outskirts of Kingston; it has since been incorporated into the city. As an estate on the outskirts of the entrepôt, Hope was most likely a staging area for Radnor coffee from whence it was transported to the docks for export. It is interesting that this was the ending point of the journeys with the coffee, rather than the wharves themselves. Although a matter of speculation, the use of this terminus may have been part of a strategy to prevent the enslaved porters from joining the crews of ships, and hence escaping slavery.

The only recorded types of punishment inflicted on the population at Radnor can also be interpreted as spatialities of control. The daily work entry for Wednesday January 23, 1822, contained the ominous passage: "masons with 5 negroes carrying stones preparing to build stocks room." This work continued through February 8; entries for this work alternately defined the space as a "stocks room" and a "stocks house." It is likely that the stone building that was under construction was a separate structure, within which movement would be harshly restricted by restraining people in stocks or pillories (Figure 33). The only overtly recorded punishment in the plantation journal utilized this facility of spatial control. On Tuesday May 3, 1825, it was recorded that "Mulatto King was detected stealing coffee; is in the stocks." It is unclear how many people were confined in this space of punishment. What is clear is that such a strategy of radically restricting spatial movement, at least on this one occasion, was used on the plantation as a means of discipline. It is likely that the building was used on more than this one occasion. Its very presence, no doubt, served to intimidate and thus discipline the population, by the threat of restricting movement by having one's hands and/or feet confined in the apparatus, and being locked in what essentially was a stone dungeon.

As Leone (1995), Orser (1988c), and Upton (1988) have recently argued, the position of structures and the division of social space on a landscape can be interpreted as expressions of power. As Foucault has noted in *Discipline and Punish* (1979), surveillance through the practice of panopticism serves as a control over both space itself and

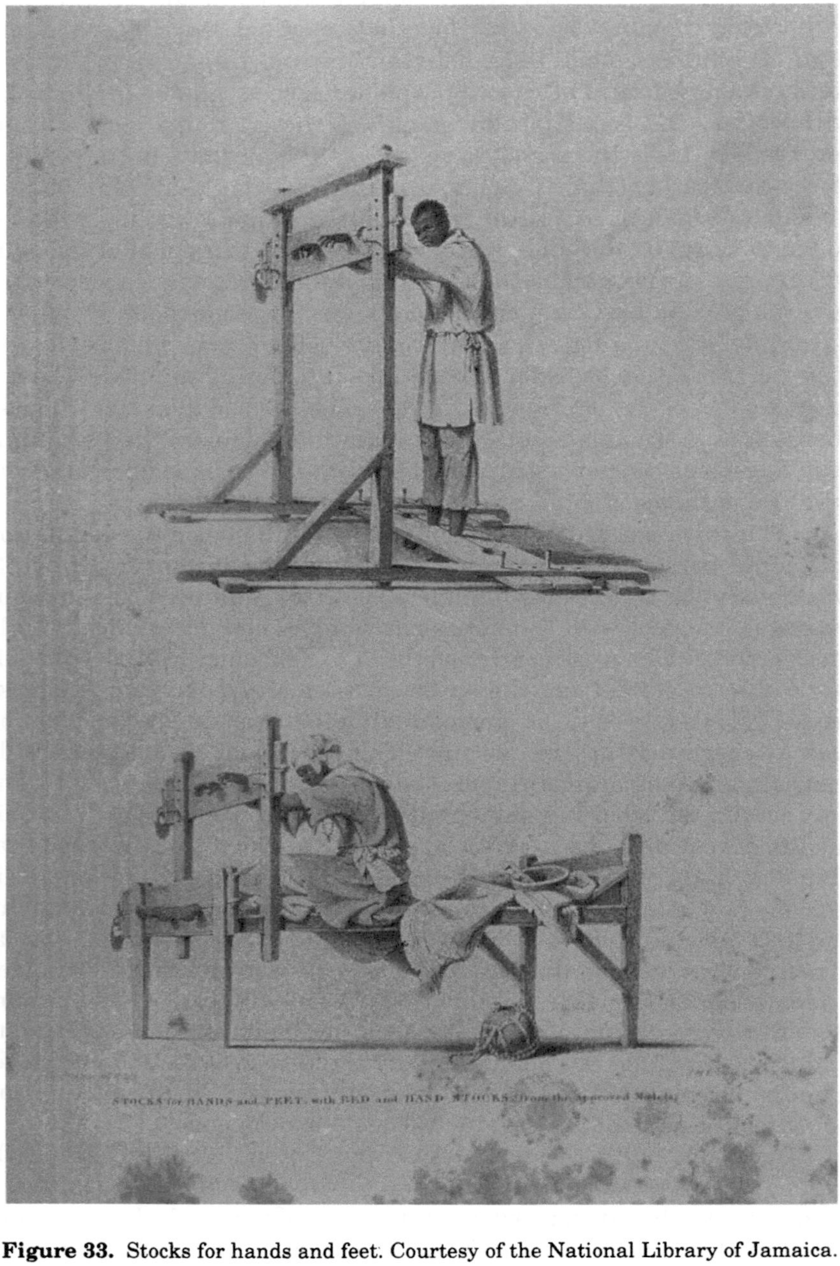

Figure 33. Stocks for hands and feet. Courtesy of the National Library of Jamaica.

movement through space. Foucault argues that panopticism is the opposite of the dungeon; rather than enclose, deprive of light and hide, panoptic surveillance through "full lighting and the eye of a supervisor captures better than darkness. . . . Visibility is a trap" (Foucault, 1979:200). Although an element in much of the architecture of capitalism, from the Georgian order to Nazi pageantry, the spatiality of surveillance as a means of social control may be nowhere better expressed than in plantation contexts.

The material record of the Yallahs plantations indicates that this means of social control was intentionally designed and exerted in the spatialities of coffee production. Interpreting the layouts of the several Yallahs estates reconstructed by plantation surveys suggests how the spatialities of surveillance were constructed and experienced during the preemancipation era.

The spatialities of surveillance may be best reconstructed on Clydesdale, given the superior preservation of the plantation buildings and the extent of the cartographic record. Using these two data sources, it is possible to create a composite map that can usefully reconstruct the spatialities of surveillance (Figure 34). Considering the plantation model as a panopticon, the overseer's house would have served as the central point of surveillance, analogous to the central guard tower of the true Benthamite panopticon. The Clydesdale overseer's house featured two surveillance positions; the first was the entrance door to the domestic quarters, which would most likely have had a small landing at the top of a wooden stair. From this point, the overseer could monitor the slave village, which was located uphill, within the viewscape of this position. The path from the village to both the coffee fields and the industrial works passed directly by this point. Thus, without leaving the confines of his house, the overseer could survey the domestic quarters of the workers, and watch them as they walked from their houses to their work.

The second surveillance point was the veranda of the overseer's house. The coffee works and barbecues were located downhill and within the viewscape of this vantage point. Thus, from the comfort of his veranda, the overseer could supervise the coffee works, and any activity occurring on the barbecues. During those times when the overseer wanted to exert the greatest measure of control over the workers, he could practice panoptic surveillance over the population. Equally crucial to this method of social control is the spatiality of the observed. By locating the overseer's house in such a way that the overseer could be surveying the village and works from the veranda, or even by gazing out of one of the house's windows, the workers could

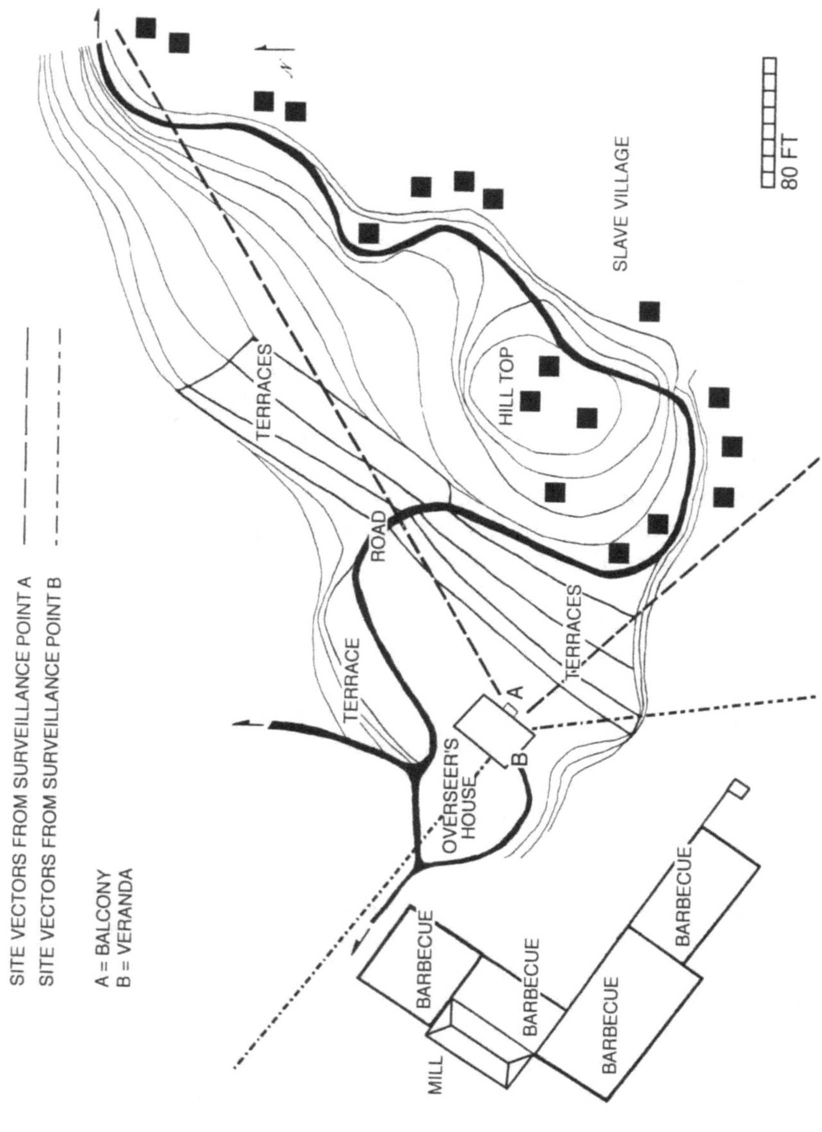

Figure 34. Composite surveillance map of Clydesdale.

never be entirely sure whether they were being watched. The purpose behind the construction of this spatiality was the construction of discipline in the work force; the logic of the panopticon dictated that the workers would cooperate if they thought they were being watched.

Similar spatialities may have been constructed at other plantations. However, the composite map for Sherwood Forest is more conjectural than for Clydesdale, as the location of the village is not recorded in the cartographic record. At Sherwood Forest (Figure 35), the original overseer's house was constructed very similarly to the Clydesdale house. From the side entrance to the house, the overseer had a vantage point from which he could survey the flats below the coffee works. These flats, now anthurium beds, may have been the location of the slave village; the elder Dr. Deichman, prior to his death, related to me that he had found numerous artifacts in that area. If this was the location of the village, the overseer's house would have served as a surveillance point from which the domestic lives of the workers could be monitored. The veranda of this house provided a vantage point from which the overseer could monitor the barbecues. The surveillance of production may have been even more intense at Sherwood Forest than it was at Clydesdale, given that the mill machinery was located in the same building as the domestic space of the overseer. At Sherwood, the overseer could directly monitor the processing of coffee from his room or veranda; it is possible that he also could have surveyed the domestic lives of the enslaved.

The evidence for Chesterfield indicates a similar spatiality of surveillance. As depicted in the composite map, the overseer at this plantation could monitor the barbecues and the mills from his house. In addition, the overseer on this plantation could see the coffee fields from the house, albeit they were located across the Yallahs River. There is no evidence to suggest that the overseer's house ever featured a veranda, an obvious surveillance feature found at both Clydesdale and Sherwood Forest. Nevertheless, from the vantage point of this house, an overseer could monitor most of the productive activities occurring on the plantation. In this case, the workers could never be certain when they were being watched; the overseer's house and the great house would both be visible from the coffee fields, the barbecues, and the mill complex.

Spatialities of Resistance

Despite the intensity of the spatialities of control on these coffee plantations, the workers exercised spatialities of resistance, which they

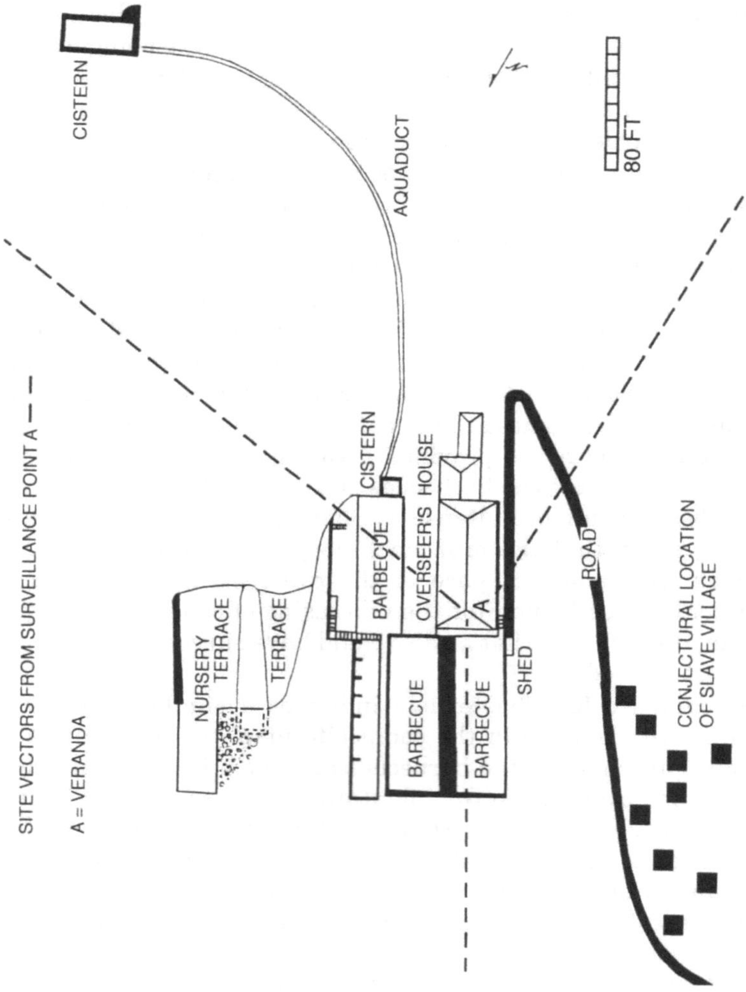

Figure 35. Composite surveillance map of Sherwood Forest.

themselves defined and controlled in dialectical contradiction to the spatialities of control being imposed on them from the planters. The documentary record of Radnor again serves as the best source for the interpretation of these spatial phenomena.

One way by which the workers could resist the spatialities of control was by devising ways to avoid becoming the disciplined and docile work force that the planters demanded. The spatialities of plantation production provided at least one category of space within which the workers could avoid coerced labor: the hospital and yaws house. It was typical for Jamaican planters to construct a hospital on the plantation grounds to serve the medical needs of the African Jamaican population; in 1826, the Jamaica Assembly passed a law requiring planters to clothe, house, attend, and maintain sick or infirm slaves (Sheridan, 1985:273). The plantation journal of Radnor indicates that a hospital existed on the grounds to attend to the sick. The work records for the estate indicate that between 4 and 12 people were usually in the hospital on a given day. Reporting to the hospital may have been a strategy used to avoid work, thus the slave hospital at Radnor may have provided the workers a spatiality of resistance. This is born out by several intriguing entries. Between Monday, December 15, and Monday, December 22, 1823, the number of people in the hospital jumped from 8 to 19. The following week was the Christmas holiday; no work was scheduled for Thursday, Friday, and Saturday of that week. The sudden increase in the number of people in the hospital may indicate that a number of people decided to extend their time away from labor, effectively taking a 2-week break from the labor regime.

Many plantations had an additional sick house on the grounds dedicated to the isolation of people with yaws. Yaws is a communicable disease caused by a spirochete related to, and in many ways similar to, syphilis; unlike syphilis, however, yaws is not necessarily a venereal disease, and can by transmitted by casual contact. The disease is characterized by rheumatic-like bone aches, fungous oozing pustules, and in advanced stages, by destruction of the facial bones which results in horrible disfigurement (Sheridan, 1985:83–84). Because of fear of contagion, and the gruesome appearance of those inflicted with this disease, it was common practice to quarantine yaws victims in houses removed from the rest of the population. Sheridan reports that depending on the management strategy on a given estate, yaws patients would either live as lepers in some remote corner of the estate while the disease ran its natural course, which could take years, or else be given medical attention by an African Jamaican woman familiar with traditional African remedies for the disease (Sheridan, 1985:88).

The Radnor plantation journal suggests that yaws victims were segmented from the rest of the population on the estate, although it is unclear whether they lived removed from the population, in some distant corner of the 987-acre estate. The record does indicate that this segmented spatiality of disease was yet another forum of contestation between the planters and the enslaved. From the commencement of the plantation book in 1822 through September 28, 1823, 19 people were listed under the column "Got Yaws"; their names, however, are not recorded. This represents almost 9% of the entire African Jamaican population of the estate. While I do not mean to imply that people would intentionally contract this disease to resist slave labor, it is possible that people capitalized on the isolation of yaws cases to conceal their actual medical condition. In doing so, they may have isolated themselves from the estate population out of the surveillance of the white planters, either feigning yaws or else exaggerating their condition.

The contested nature of the spatiality of yaws is revealed in the daybook entries for the fortnight between September 22 and October 5, 1823. At the beginning of this period, 19 were listed with yaws, as they had for the previous 21 months, while only 8 people were listed as being in the hospital. By Thursday of that week, the number of people in the hospital had increased to 16; on Friday the number of people had increased to 29. There is no indication why there was a sudden influx of people into the hospital; it may have been a type of labor strike against the plantation. On the following Monday, 12 of the people with yaws were transferred to the hospital where they stayed until the following Monday. On October 6, the number of people in the hospital was reduced to 9; only 7 yaws cases were reported. This number remained consistent until October 24, when the number increased to 9. The question remains: Why were 12 people suddenly removed from the yaws list and put in the hospital? It may well be the case that the plantation managers used the opportunity of having such an inordinate number of people in the hospital to reevaluate the quarantined yaws patients, and after an examination and short stay in the plantation hospital, their condition was upgraded to the point where they could rejoin the labor force. This may indicate that many of the yaws patients were using the quarantine as a strategy of resistance, and may not have been as ill as the planters were led to believe.

It is clear that beginning on October 6, 12 of the 19 yaws patients were returned to the work force. Why did this occur when it did? It is plausible that the planters used the yaws patients as a scare tactic by reintroducing them to the hospital. By bringing people with the highly contagious disease into contact with workers using the hospital as a spatiality of resistance, the planters may have hoped to frighten people out of the hospital and back to work. In either scenario, the yaws house and the hospital can be interpreted as contested spaces used to negotiate the spatiality of resistance.

Enslaved workers on Radnor used unsanctioned access to the spatiality of movement as a further expression of resistance. Such strategies are recorded in the plantation journal alternately as people absconding, being absent, or running away; all synonyms to describe movements off of the estate unauthorized by the planters. The Radnor plantation book covers the period from January 1822 through February 1826, save the period January 4 to July 5, 1823, which is missing from the book. The plantation staff recorded the names of workers who escaped from the plantation, the dates they left, and the dates they returned. This information is reproduced in Table 4. In some cases, the plantation book records short remarks about the circumstances of their return.

In all, 25 different people — about 11% of the total population or 16% of the adult population — absconded from the plantation a total of 33 times; 8 escaped twice, 17 once. The dates of escape and return for all but 8 of these incidents were recorded. Of the 25 people who managed to escape, 11 were women and 14 were men. Of these, all but 1, Phoebe, eventually returned to the plantation. Of the incidents with recorded dates of escape and return, the average time away from the plantation was 19 days. Besides Phoebe, only Trim (79 days), Flora (21 days), Little Quomin (55 days), Murray (36 days), Fox (27 days), and Matthew (43 days) were gone for a longer period than the mean. The journal records several additional instances of unsanctioned movement that cannot be quantified, as the date the people returned and/or the number of days they were absent from the plantation do not appear in the journal (Table 5).

Through the process of escape, the enslaved population of Radnor expressed control over the spatiality of movement, at least for a brief time. This unsanctioned movement was in resistance to the spatialities of control; the frequency of escape and sheer percentage of the population that chose to express this particular spatiality of resistance demonstrate that despite the consequences of punishment, movement through space was regularly exercised by at least this group in the Yallahs region.

Table 4. Runaways and Absenteeism of Known Duration at Radnor, February 1, 1822 through February 1, 1826. Entries for January 4 through July 5, 1823, Excluded; Missing from Mss.

Name	Sex	Date absconded	Date returned	Days absent	Remarks
Ruth	♀	3/17/1822	3/19/1822	2	
Dunkin	♂	3/20/1822	3/26/1822	6	
Hellen	♀	3/28/1822	4/8/1822?	11	
Betty	♀	4/15/1822	4/26/1822	11	
Sylvia	♀	5/17/1822	6/4/1822	18	
Trim	♂	5/23/1822	8/10/1822	79	
Josey	♂	8/22/1822	9/9/1822	18	
Little Quomin	♂	8/19/1824	8/23/1824	4	
Precilla	♀	8/30/1824	9/4/1824	5	
Lucy	♀	12/20/1824	12/22/1824	2	
Flora	♀	12/9/1824	12/30/1824	21	
Lucy	♀	12/9/1824	12/21/1824	12	
Precilla	♀	12/9/1824	12/21/1824	12	
Rainsford	♂	2/9/1825	2/16/1825	7	
Cork	♂	2/9/1825	2/20/1825	11	"Got Carpenter George to beg for him"
Sp. King	♂	2/9/1825	2/25/1825	16	
Matthew	♂	2/9/1825	2/25/1825	16	
Little Quomin	♂	2/9/1825	4/5/1825	55	Caught; detained in St. George workhouse
Juliet	♀	2/10/1825	2/28/1825	18	
Betty	♀	2/12/1825	3/1/1825	17	Robbed Thomas Kelley
Sp. King	♂	8/15/1825	9/1/1825	17	"Caught and brought home"
Fox	♂	10/13/1825	11/9/1825	27	
Murray	♂	10/24/1825	11/29/1825	36	
Matthew	♂	12/5/1825	1/17/1826	43	"Brought home by Mulatto King"
Quomin	♂	1/9/1826	1/16/1826	7	

Source: Radnor plantation book.

Table 5. Runaways and Absenteeism of Unknown Duration at Radnor, February 1, 1822 through February 1, 1826. Entries for January 4 through July 5, 1823, Excluded; Missing from Mss.

Name	Sex	Date absconded	Date returned	Days absent	Remarks
Phoebe	♀	5/7/1821	did not	??	Only successful runaway
Hellen	♀	??	3/26/1822	??	
November	♂	4/10/1822	??	??	
Dunkin	♂	4/26/1822	??	??	
Fox	♂	5/25/1822	??	??	
Grace	♀	5/31/1822	??	??	
Cato	♂	6/1/1822	??	??	
Amanda	♀	8/9/1822	??	??	

Source: Radnor plantation book.

CONCLUSION

Between 1790 and 1834 the Yallahs region experienced profound changes in the organization of space. Prior to this period, the upland river valleys were wilderness areas, nominally owned by patentees, but sparsely populated. With the introduction of coffee production, the region was transformed into commodity-producing space. The landscape was carved up and subdivided into parcels, which were conveyed as private property to coffee speculators. The closing decade of the eighteenth century and the opening decades of the nineteenth century witnessed a further redefinition of space, as the vacant landscape depicted in the early maps of the region was transformed into a number of working plantations. Coffee fields were carved out of the forest, terraces were carved out of the hillsides, massive structures were built to accommodate overseers and planters, rivers were diverted to operate machinery and act as sewers for coffee waste. Hundreds of people were transported into the mountains to work on the estates. Dozens of villages were produced, houses built, and provision grounds cleared and planted. A new variation of productive space appeared on the Jamaican landscape dependent on slave labor and a newly introduced labor regime.

The spatialities that existed under the slave regime were an elemental part of the construction of racial and class relations in the Yallahs region. Through these spatialities the planter class attempted to maintain control over the labor power of the enslaved, while the African Jamaican population used their own definitions of space to both proactively and reactively resist the spatial impositions of the planters. Beginning in 1834, however, the conditions under which these spatialities were negotiated changed. With the abolition of slavery, the Yallahs region entered a new phase of spatial restructuring. Chapter 7 explores the spatial dimensions of this postemancipation era in the Yallahs drainage.

7

Postemancipation Developments

1834–1865

Free negroes are found to act differently from other free men; not because they differ from others in character, but because their circumstances are different; and just in proportion as they are brought within the reach of those motives by which Europeans are governed, will their conduct resemble that of the natives of Europe.
— Henry George Grey to the Deputation of the Standing Committee of West India Planters, 1833

INTRODUCTION

The relationship between labor and space in Jamaica changed dramatically in the years between 1834 and 1865. Two developments in the 1830s radically changed the relationship between laborers and the planters, and encouraged the redefinition of social and material spaces throughout Jamaica. In 1834, slavery was abolished in the British West Indies. In that year, a new labor system, known as apprenticeship, replaced slavery. In 1838, the apprenticeship system was abolished in Jamaica, and the relationship between the African Jamaican population and the European elites entered a new phase, mediated by wage labor rather than the bonds of slavery. Jamaica's transition to wage labor was marked by occasionally violent episodes of unrest. Among the most dramatic was the Morant Bay Rebellion of 1865, an uprising that occurred in the shadow of Blue Mountain Peak and that would result in a complete restructuring of Jamaica's government. In the wake of the Morant Bay Rebellion, the Jamaican legislature was suspended, and the affairs of the island brought under the direct control of the Westminster Parliament (Heuman, 1994). This episode is often used as a bracket within the historiography of Jamaica, and serves as the effective terminal point of this study.

British elites designed the apprenticeship system to be a transitional step toward wage labor. Under the apprenticeship system, able-bodied workers remained attached to the estates on which they had been enslaved. While enslaved the workers had been valuated as capital; with the advent of freedom, the planters demanded compensation from Parliament for the emancipated workers, whom the planters had defined as their private property. In determining how much the masters would be compensated for the value of the enslaved laborers, in 1834 African Jamaican people were classified by relative value within the apprenticeship system. "Praedial attached" apprentices were those who were employed in agricultural pursuits on land belonging to their owners; "praedial unattached" apprentices were agricultural workers who labored on land owned by someone other than their owner; "nonpraedial" apprentices were those who were engaged in other kinds of employment, e.g., as domestics (Butler, 1995:30). Children under the age of 6 were emancipated unconditionally (Green, 1976:134).

Under the Emancipation Act, praedial apprentices were to serve a period of coerced labor extraction of 6 years before they were fully emancipated; nonpraedial apprentices were to be freed after 4 years. Under the terms of apprenticeship law, the former slaves were required to work for 45 hours for the planters per week. Beyond this weekly period, the apprentices were free to use their labor power as they saw fit, either working their provision grounds, or else selling their labor power to the planters for wages. It was hoped by the British who concocted this plan that apprenticeship would acclimatize the former slaves to the rigors of wage labor, smoothing the transition from slavery to competitive labor (Holt, 1992:56–57). In large measure a result of the widespread resistance to the apprenticeship system by both the apprentices and the antislavery lobby in Britain, the apprenticeship system was terminated 2 years ahead of schedule. Jamaican apprenticeship was abolished on August 1, 1838. At that time the formerly enslaved African population entered a state that has come to be known as "full freedom" (Green, 1976:129–130; Wilmot, 1984). It should be noted that while the West Indian planters were compensated a total of £20,000,000 for the value of their slaves, the enslaved themselves never received compensation from the British government.

These developments, which can be interpreted as part of the process of crisis-driven restructuring of the Jamaican political economy, resulted in significant redefinitions of spatial meanings and relationships. In this chapter, I will discuss how these transformations were manifested in the Yallahs coffee-producing region.

THE HISTORICAL DEVELOPMENT OF COFFEE PRODUCTION, 1834–1865

The restructuring of labor under and after apprenticeship affected coffee production in the Yallahs region in several ways. The recording of coffee crops in the Accounts Produce demonstrates some unusual trends for the years surrounding the apprenticeship period. The last recorded entry for Mt. Charles appeared in 1839, the year immediately following full emancipation. Chesterfield disappeared from the Accounts Produce in 1830. The last entry for Radnor was made in 1838. Although appearing only three times in the record, the last entry for Cocoa Walk was made in 1839. Of the 10 estates whose production I traced through the Accounts Produce, only 4 reported crops after 1839. Although Mavis Bank had been abandoned as a coffee estate well prior to emancipation, there is no direct evidence to suggest that the other properties were abandoned, despite the fact that they do not appear in the Accounts Produce. Indeed, in a report filed in Parliament in England, it was reported that between 1832 and 1848 only 1 coffee plantation, Washington, was abandoned in the parish of Port Royal; 4 (Burness and Bothwell plantation; Friendship Hall and Mt. Pleasant plantation; Clyde Side plantation; Orange Park plantation) were abandoned in St. David. In contrast, 111 coffee plantations were abandoned in the neighboring parish of St. Andrew, and 109 in the parish of Manchester. In total, 465 coffee plantations, comprising 188,400 acres, were reportedly abandoned throughout the island between 1832 and the date of the report in 1848 (PP, 1848:23/3, 225–230).

From the planters' perspective, the coffee industry entered a crisis following emancipation, primarily based on the increasing difficulty planters had controlling labor. For example, in a sworn statement before a special committee of the Jamaica Assembly appointed in 1847, William George Lowe, who owned coffee properties in the contiguous parishes of Port Royal, St. George, and St. Andrew, reported difficulty in obtaining labor to cultivate his estates. When asked about the condition of his properties, Lowe reported:

> My property, Green Hill, in St. George's, I have been obliged to abandon for want of labour. When the coffee was ripe on the trees, I was unable to find labourers to gather it in. In 1833 there were 52 labourers attached to the property; more than two-thirds of them have purchased or leased lands, and become independent settlers; the other property, Violet Bank, I continue to cultivate, but not now to advantage. (PP, 1848:23/3, 163)

During the same set of hearings concerning the condition of sugar and coffee cultivation in Jamaica, Hugh Fraser Leslie reported that in that year, 1847, he was in possession of several coffee and sugar plantations, including several in the Yallahs region: Sherwood Forest, Arntully, Eccleston, Brook Lodge, and Belle Claire. In addition, Leslie was also in possession of Petersfield, Newfield, Munts, Grove, and Leith, all coffee plantations, and Delve, a sugar estate. He reported that prior to emancipation, a total of 1268 workers were attached to these plantations. By 1847, there were fewer than 400 (PP, 1848:23/3, 167). Leslie complained that he had a difficult time organizing and controlling labor on these estates, which hampered his efforts at efficient estate management:

> From the indisposition of the people to work during the Christmas holidays, a great sacrifice of property takes place from the loss of coffee, which falls to the ground from want of hands to pick it during a period of two or three weeks, when no wages, however extravagant, can obtain their labour, this too being the period when the ripening of the coffee is the most general. A further loss takes place in not being able to pick off what is termed the *one* coffee at the commencement and close of each crop. Another difficulty is the refusal of the people to work on Fridays and Saturdays, even when the coffee is ripe on the field. (PP, 1848:23/3, 167)

David McLean, the owner of Middleton plantation and manager of Sheffield and Manheim, all coffee estates in St. David, corroborated that planters operating near the Yallahs region were experiencing difficulty in controlling postemancipation labor on coffee plantations. In his testimony to the same committee, McLean stated:

> I pay at the rate of 1s to 1s 6d for day labour, and at the rate of 12s per acre for cleaning, and 14s per acre for pruning, for picking from the field, 1s per bushel, and 1s per 100 lbs for picking for the market. I do not receive half of a fair day's work for the money paid; the people work only four days in the week, even when the coffee is ripe in the fields, and at Christmas time, when the coffee is falling from the tress, they will not work on any terms, even if I offer them a guinea a bushel; and I am satisfied that one-sixth of the crop is lost in consequence of these difficulties of labour. (PP, 1848:23/3, 164)

McLean may have been an exceptionally poor manager of wage laborers. Holt reports that a "Mr. MacLean," coroner for St. David, was killed in a political riot in 1851. Fifty-five people were later tried for rioting; eight were indicted for MacLean's murder (Holt, 1992:232). It seems very likely that this was David McLean, as an anonymous dispatch from a Jamaican magistrate, dated 1857, described Middleton as a coffee estate in St. David "on which the late coroner resided for several years" (CO, 137:334, 34).

Discontent with the system of apprenticeship appeared nearly from its inception. This discontent was manifested in the Yallahs region as elsewhere in Jamaica. In 1835, the first year of apprenticeship, Frederick French, overseer of Green Valley, reported to a committee of the Jamaica Assembly that he had experienced difficulty in either compelling or convincing the apprentices on that estate to spend their free time picking coffee. From his perspective, the plantation ran better under the regime of slavery. When asked whether the apprentices were working "as cheerfully and as well as they did previous to the 1st of August," i.e., the beginning of the apprenticeship system, French replied:

> Most unquestionably not. While slaves, they were contented and industrious, and, in this neighborhood, remarkable for their good conduct; but now, with the same allowances as formerly, and additional time allowed them by law, they are lazy, insubordinate and extremely insolent; the most civil question put to any of them is invariably answered impertinently, and generally accompanied with threat that they will go to the magistrate. (PP, 1835:50, 100)

Clearly, as the African Jamaican population began to explore the social dynamics of their newly acquired status as apprentices, they were freer to express their opinions of and attitudes toward the plantation overseers. This change in attitude was reinforced by the Emancipation Act, which provided that special magistrates would be appointed to protect the apprentices from many of the more extreme forms of punishment that had been meted out to them while under the system of slavery.

The quest for overt expressions of independence on the part of the formerly enslaved workers frustrated the attempts of the planters in the Yallahs region to conduct coffee cultivation as they saw fit. French reported that at Green Valley, he had a very difficult time convincing the apprenticed workers to sell their labor power to the estate during their free time. When asked if he knew of any example of apprentices declining to work for wages when so requested, French replied:

> Yes, I have been myself in the habit of hiring the people on their own days, and paying fair wages. . . . This year there is near double the quantity of coffee on the trees, and considerably riper than at the same period last year. I have several times begged them to pick coffee in their own time . . . but could not get one to do so this year, nor to pick near the quantity for the property that they ought to do. (PP, 1835:50, 100)

The frustrations expressed by white elites like Leslie and French may have overstated the reality of the situation immediately following the enactment of the apprenticeship system. Official reports sent to Parliament from the special magistrates assigned to the Yallahs region, Patrick Dunne and Henry Kent, minimized the conflict between labor and management. Table 6 lists the estates over which each of these two men had jurisdiction. In June 1835, Dunne reported that he considered

Table 6. Estates in the Jurisdiction of Special Magistrates Dunne and Kent, 1835

Henry Kent		Patrick Dunne	
Chester Vale	Farm Hill	Friendship	Mt. Sinai
Clydesdale	Hibernia	New Battle	Norris
Industry	Whitfield Hall	Ayton	Aeolus Valley
Resource	Abbey Green	Sheffield	Lloyd's
Pleasant Hill	Epping Farm	Ultimatum	Windsor Castle
Mt. Hybla	Windsor Lodge	Friendship Hall	Albion
Hall's Delight	Mahogoney Vale	Carrick Hill	Richmond
Mt. Faraway	Mt. Charles	Sherwood Forest	Woburn Lawn
Old England	Orchard	Minto	River Head
Robertsfield	Penhill	Grove	Windsor Forest
Green Valley	Westphalia	Moy Hall	Belle Claire
Strawberry Hill	Bemah [sic]	Morha	Fair Prospect
Mt. Teviot		Clifton	Arntully
Bryant's Hill		Mt. Pleasant	Manheim
Sheldon		Radnor	Hermitage
Penlyne Castle		Cocoa Walk	

"the negroes in my district as well disposed, from their having on every occasion of my applying to them entered into arrangements to work during crop seven hours per day beyond the time fixed by law" (PP, 1835:50, 254). Kent confirmed that a similar condition existed on the estates in his district: "the negroes in general throughout this district seem well disposed, and I think perform their labor willingly" (PP, 1835:50, 258–259).

The contradictions between the reports made by estate agents and those made by the special magistrates may indicate that the African Jamaican population held the magistrates in higher regard than they did the elites who managed the estates. This interpretation is corroborated by an anecdote reported by Holt (1992:64). According to Holt, when an unnamed woman refused to work on Pusey Hall estate, the overseer threatened to call the special magistrate to inflict a punishment for indolence. Rather than wait for a decision, the woman took a shortcut to the magistrate's office, arriving before the overseer, to give her side of the story. Although planters may have emphasized the difficulties they were experiencing in attempts to influence Parliament, instances such as this indicate that the laboring population respected the magistrates, and attempted to use the newly imposed judicial system to their advantage.

Resistance to the apprenticeship system was expressed in several ways. Refusing to work on the estates for wages during the busiest crop season was one such strategy. A similar strategy was to refuse to work at all, or else to refuse to conform to the regimen required by the overseers. Henry Kent reported that this latter form of resistance was more common among women than men. In his 1835 report to Parliament, Kent commented: "women give much more trouble than the male apprentices, from an unwillingness to go to the field at shell blow" (PP, 1835:50, 259). Holt argues that this trend was common throughout Jamaica (Holt, 1992:64).

The tendency to resist the terms of apprenticeship is reflected in a table appearing in the British Parliamentary Papers in 1836 (PP, 1836:48, 219). During the first year of the apprenticeship system, Parliament required the special magistrates to quantify the "offenses" perpetrated by the apprenticed laborers in their respective districts, and to report on the kind and quantity of punishments they had meted out. Between August 1834 and July 1835, Patrick Dunne reportedly inflicted punishments on 562 people — 357 men and 205 women. The most common offenses were overt acts of resistance to the new labor regime: 279 people were found guilty of "neglect of duty," 144 for "disobedience," 52 for "indolence," 21 for running away, 60 for theft, and 6 for cutting or maiming cattle. In the 3 months between May and July 1835, Henry Kent reported that he had punished 63 males and 32 females: 39 for neglect of duty, 27 for disobedience, 6 for indolence, 3 for running away, and 20 for theft. While it is difficult to say whether the thefts should be interpreted as a form of resistance, as the plaintiffs were not identified, the other categories of offense ("neglect of duty," "indolence," "disobedience," "runaway") clearly define behaviors that challenged the plantation elites (Table 7).

Punishments for these offenses, which should really be considered as expressions of refusal to conform to the demands of apprenticed labor,

Table 7. Offenses Heard by Special Magistrates, August 1834–August 1835

"Offenses"	Henry Kent (May–July 1835)	Patrick Dunne (August 1834–August 1835)	Total for Jamaica (August 1834–August 1835)
Theft	20	60	2,837
Runaways	3	21	1,805
Neglect of duty	39	279	11,855
Disobedience	27	144	6,024
Cutting/maiming cattle	0	6	322
Indolence	6	52	2,552
Total	95	562	25,395

Source: PP, 1836:48, 219.

ranged from floggings and switchings to imprisonment and solitary confinement. The most common form of punishment meted out by both Dunne and Kent was to compel the laborers to make up ("repay") missed time. In addition, each magistrate sentenced one individual to the treadmill, a device newly introduced to Jamaica. First introduced to the English prisons in 1818, the treadmill was designed as a mechanism to instill discipline in the work force — both the discipline of routinized labor and the discipline of punishment for not conforming to the demands of wage labor (Table 8; see Figure 36). According to Holt, in Jamaica, the treadmill was used more as an instrument of torture than a device for creating labor discipline (Holt, 1992:106).

Although the laborers entered apprenticeship with the knowledge that it would be a temporary stage before the advent of full freedom, many chose to resist the system by purchasing their freedom. The Emancipation Act allowed a provision by which the special magistrates could determine the value of apprentices as chattel; if the apprenticed African Jamaicans could acquire the valuated sum and turn it over to the management of the estate to which they were attached, they were guaranteed freedom. Several laborers attached to estates in the Yallahs region chose to purchase their freedom in this manner. In all, 29 African Jamaicans from Yallahs estates were reported to have purchased their freedom (Table 9). This strategy seemed to work better for women than men; of the 29 people who acquired full freedom by compensating their masters, 23 were women. This may reflect the difficulty men had in accumulating enough wealth to purchase their freedom, as the planters tended to demand a higher price from men for their full emancipation. In the Yallahs region, the average price paid by the 23 women who

Table 8. Punishments Meted out by Special Magistrates, August 1834–August 1835

"Punishments"	Henry Kent (May–July 1835)	Patrick Dunne (August 1834–August 1835)	Total for Jamaica (August 1834–August 1835)
Flogging	28	83	7125
Imprisonment	3	6	1249
Treadmill	1	1	1176
Penal gang	0	1	2941
Repay time	40	360	9433
Solitary confinement	12	90	2886
Switching	11	21	585
Total	95	562	25,395

Source: PP, 1836:48, 219.

Figure 36. A workhouse treadmill scene in Jamaica. Courtesy of the National Library of Jamaica.

Table 9. Valuations of Apprentices, on Yallahs Estates, 1835–1837

Name	Sex	Class	Estate	Trade	Date	Sum paid	Unpaid	Remarks
James Rhoden	M	Praedial	Epping Farm	Mechanic	1835	£57.3.4		
Sarah Richards	F	Praedial	Epping Farm	nd	1835	£49.4.6		
Joe Leslie	M	Nonpraedial	Green Valley	nd	1835		£51.11.0	
Mary Palmer	F	Praedial	Flamstead	nd	1835		£56.0.0	
Elizabeth Smith	F	Nonpraedial	Whitfield Hall	nd	1835	£31.2.5		
Margaret Walker	F	Nonpraedial	Whitfield Hall	nd	1835	£33.4.6		
Mary O'Hanlon	F	Nonpraedial	Whitfield Hall	nd	1835	£11.1.5		
Mary Anne Parker	F	Nonpraedial	Westphalia	nd	1835	£39.7.3		
Sarah Lee	F	Nonpraedial	Whitfield Hall	nd	1835		£35.8.8	
John Phipps	M	Nonpraedial	Radnor	nd	1836	£48.17.9¼		
Maria Rock	F	Nonpraedial	Mt. Teviot	nd	1836	£34.6.8		
Rosanna Gardner	F	Praedial	Chester Vale	nd	1836	£20.12.3		
Ann Devaney	F	Praedial	Green Valley	nd	1836	£26.13.11¾		
Henry Wilkie	M	Nonpraedial	nd (Robert Law)	nd	1836	£28.19.9½		
Violet Wilkie	F	Nonpraedial	nd (Robert Law)	nd	1836	£20.0.0		
Francis Walker	M	Praedial	Belle Claire	nd	1836		£35.6.8	
Eleanor Walker	F	Praedial	Belle Claire	nd	1836		£47.2.4	
Andrew Cambell	M	Praedial	Belle Claire	Carpenter	1836		£84.10.0	
Brown Hall	M	Praedial	nd (William Rae)	nd	1836		£44.16.0	
Michael Francis	M	Praedial	Cocoa Walk	nd	1836		£10.17.9½	
John Lowe	M	Praedial	nd (Wm. George Lowe)	nd	1836		£55.3.8	
George	M	Praedial	Orchard	nd	1836		£72.8.4	
Eliza Smith	F	Nonpraedial	Whitfield Hall	nd	1836	£31.2.5		
Margaret Walker	F	Nonpraedial	Whitfield Hall	nd	1836	£33.4.6		
Mary O'Hanlan	F	Nonpraedial	Whitfield Hall	nd	1836	£11.1.5		
Sarah Leigh	F	Nonpraedial	Whitfield Hall	nd	1836	£35.8.8		
Mary Ann Parke	F	Nonpraedial	Westphalia	nd	1836	£39.7.3		
Sarah Richards	F	Praedieal	Epping Farm	nd	1836	£49.4.6		
Rossannah Gardner	F	Praedial	Chester Vale	nd	1836	£20.12.3		

Mary Jones	F	Nonpraedial	Mocho	nd	1836	£26.13.4		
Eleanor Richards	F	Praedial	Mocho	nd	1836	£43.16.8		
Agnes McGowan	F	Nonpraedial	Green Valley	nd	1837	£22.6.3		
Eleanor Yates	F	Nonpraedial	Flamstead	nd	1837	£20.0.0		
John Walker	M	Praedial	Chesterfield	Mason	1837	£78.1.8		
Mary Henry	F	Nonpraedial	Cocoa Walk	nd	1837	£32.0.0		
Mary Wright	F	Praedial	Windsor Forest	nd	1837	£41.0.0		
N. Forbes	M	Praedial	Middleton	nd	1837		£60.0.0	
J. Rose	M	Praedial	Cocoa Walk	Cooper/ carpenter	1837	£102.6.8		
Johanna McLarty	F	Praedial	Chester Vale	nd	1837	£22.10.4		
Sarah Rattray	F	Praedial	Clydesdale	nd	1837		£35.17.10	Has not the funds
Sarah Barnett	F	Praedial	Abbey Green	nd	1837		£42.0.0	Has not the money
Phoebe Duffus	F	Praedial	Industry	nd	1837		£44.0.0	Too high
Elliston Walker	M	Praedial	Chester Vale	nd	1837	£32.0.0		
Mary Ann Purvell	F	Praedial	Chester Vale	nd	1837	£15.6.8		

Source: PP, 1836:48; 1837:53; 1838:49.

secured their freedom between 1835 and 1837 was £29.11.0; the six men paid an average of £57.18.2. Plantation artisans, all of whom were men, were more highly valued by the plantations as they were skilled laborers; they were also the people who may have been more likely to have access to cash, as they had readily marketable skills. Of the six Yallahs men who purchased their freedom during the apprenticeship period, three were artisans (one mechanic, one cooper/carpenter, and one mason). When these three plantation artisans are factored out, the average price paid by the nonskilled male laborers was £36.12.6, still better than £7 — or 25% — more than the women paid on average.

Coffee production declined in Jamaica and the Yallahs valley during the period of the apprenticeship. According to the records kept by Parliament, and reproduced in the *Jamaica Almanac,* the crop for 1833, the year that Parliament adopted the Emancipation Act, was the smallest since 1798. Although production rebounded slightly in the following year, the Parliamentary records indicate that production fell dramatically again in 1835 and 1836. The information published in the *Jamaica Almanac* indicates that this trend continued throughout the period of the apprenticeship. The figures for the gross amount of coffee exported from Jamaica, and reportedly produced on several Yallahs estates, are reproduced in Table 10.

According to these data, the average crop exported from Jamaica in the 5 years immediately previous to the implementation of the apprenticeship system (1830–34) was 16.7 million pounds by weight. In the 5 years following apprenticeship (1835–1839), including the crops produced in part or in whole by apprenticed labor, the average crop dropped to 11.1 million pounds by weight, a decrease of nearly 34%. The Yallahs plantations for which records exist reflect this trend. Each of the plantations for which statistics exist reported declining crops in the half-decade following the Emancipation Act. The larger estates seem to have had the most dramatic falls in production. The average crop reported by the management of Green Valley, which reported a crop of over 106,000 pounds in 1830, dropped 48.1% between the two half-decades. The average coffee crop on Radnor, another estate which reported a crop of greater than 100,000 pounds in 1830, dropped an astounding 52.9% during this period. Small estates, like Mt. Charles, Whitfield Hall, and Chester Vale, each demonstrated a significant drop in production, but the averages were more on par with the islandwide decrease. The average decrease on these three estates was 37.3, 39.1, and 18.8%, respectively (see Table 10).

After the system of apprenticeship ended, coffee production continued to decline throughout Jamaica. The *Jamaica Almanac* reported

Table 10. Coffee Production on Five Yallahs Estates, with the Total Exported from Jamaica, 1830–1839

Year	Mt. Charles	Whitfield Hall	Green Valley	Radnor	Chester Vale	Total for Jamaica
1830	50,524	39,198	106,262	102,443	69,160	22,256,950
1831	38,787	37,686	115,807	91,570	41,652	14,055,350
1832	28,268	16,393	66,216	42,664	57,227	19,815,010
1833	28,209	34,358	101,108	nd	nd	9,866,060
1834	44,346	15,052	nd	nd	81,274	17,725,731
Average 1830–34	38,026	28,538	97,349	78,893	62,329	16,743,820
1835	32,254	33,510	nd	70,572	84,428	10,593,018
1836	27,702	10,473	54,631	16,942	22,097	13,446,053
1837	16,680	12,780	42,195	30,786	45,295	8,955,178
1838	26,410	13,253	60,949	30,215	nd	13,551,795
1839	16,124	16,906	44,408	nd	nd	8,897,421
Average 1835–39	23,834	17,385	50,546	37,129	50,607	11,088,693
% decrease, 1830–34/1835–39	37.3%	39.1%	48.1%	52.9%	18.8%	33.8%

exports through the year 1845, the nadir in coffee production. In that year, a mere 5 million pounds of coffee was exported from the island; the largest coffee crop reported, in 1814, was more than 34 million pounds. If we consider these two extreme years, the annual coffee crop declined by more than 84% between its peak and its nadir. The average coffee crop in the first half-decade of the 1840s (1841–1845) averaged a mere 6.6 million pounds exported per annum (Table 11). Compared with the average crop exported for 1830–1834, the final half-decade during which slave labor was employed, this represents a 60.5% decrease in coffee exported from the island. The primary documents seem to indicate that the 1840s was a time of upheaval for the Yallahs coffee industry. Only 3 of the 10 estates traced through the Accounts Produce reported crops between 1841 and 1845: Whitfield Hall, Chester Vale, and Clydesdale. Of these 3, the record for Clydesdale terminates in 1843; there are no production records for this estate following this date (see Table 11).

Both Whitfield Hall and Chester Vale demonstrate declining coffee production through this period in line with the decrease in coffee production for the island as a whole. When compared with the average crop in the half-decade immediately prior to apprenticeship, the average crop on Whitfield Hall in the 1841–1845 period reflects a 56.2% decrease in production. On Chester Vale, the decrease was 52.3%. Only Clydesdale reported a relatively modest decrease in production of 18.9%, well below the islandwide decrease of 60.5%. However, as Clydesdale disappeared from the historical documents during this half-decade, it seems plausible that this estate, too, experienced economic upheaval. The historical documents seem to indicate that the Yallahs region experienced a particularly bad year in 1842, as production on each of the

Table 11. Coffee Production on Three Yallahs Estates, with the Total Exported from Jamaica, 1841–1845

Date	Whitfield Hall	Chester Vale	Clydesdale	Total for Jamaica
1841	16,348	49,069	39,246	6,438,370
1842	10,859	16,022	8,028	7,084,914
1843	15,013	40,428	25,034	7,367,113
1844	12,208	23,827	nd	7,148,775
1845	8,056	19,496	nd	5,021,209
Average 1841–45	12,499	29,769	24,103	6,612,077
% decrease from 1830–34	56.2%	52.3%	18.9%	60.5%

Sources: Accounts Produce; *Jamaica Almanac for 1846*.

estates listed in Table 11 recorded a decline in production from 1841, although Jamaica as a whole reported an increased crop.

Coffee production fell throughout the postemancipation period. By the late 1850s, many coffee plantations in the Yallahs region had ceased production and were abandoned. In an anonymous dispatch dated February 1857, a Jamaican magistrate reported that in the parish of St. David, there were only "three out of eleven estates remaining" (CO, 137:334, 24). While this number may have only indicated sugar plantations, as the word "estate" was generally used only in describing such enterprises, the number of operating coffee plantations throughout the parish was surely down. In 1852, it was reported to Parliament that only seven coffee plantations were operating in St. David, although there were more, a total of 40, operating in Port Royal (PP, 1854:43, 55). This was a significant decline in the number of operating estates in the two parishes. In 1836, it was reported to Parliament that 21 coffee plantations existed in St. David, while 51 were operating in Port Royal (PP, 1836:48, 242). In the 1850s, only Sherwood Forest, Arntully, Chester Vale, and Green Valley reported production; by the 1860s, the only Yallahs coffee plantation to appear in the Accounts Produce was Green Valley, which remained documented for the entire run of the Accounts Produce, which ceased recording crop accounts in 1867.

Declining coffee production can in part be explained by the refusal of the African Jamaican population to develop into a rural proletariat dependent on wages paid for work on the estates. As early as 1839, the first year of full freedom, the special magistrates reported that the estates were experiencing difficulty in maintaining a labor force. In discussing the state of agriculture in the Port Royal section of the Yallahs region in February 1839, only 6 months after the end of the apprenticeship system, Henry Kent complained that "few people are employed on the estates as compared with the number of apprentices they had on the 31st of July . . . there is a clear falling off in labor and I regret to say it is now eight o'clock before they get to the fields" (CO, 137:242, 238). The situation for the planters in St. David was no different. In a similar report, Kent noted that in the upper St. David district, which included estates in the Yallahs region, "few people are at work on the estates" (CO, 137:242, 240). Kent compiled labor statistics comparing the number of laborers at work on the various estates in 1839 with the number of apprentices attached to the estates in 1838. These are reproduced in Table 12.

Kent attributed the labor difficulties to a lack of available capital on the part of the planters, and the increasing economic independence

Table 12. Comparison of Number of Laborers in 1839 and Number of Apprentices in 1838

Estate	Apprentices, Jul. 31, 1838	Average no. of workers, Jan.–Feb. 1839
Sherwood Forest	103	51
Radnor	165	50
Minto	69	16
New Battle	63	34
Mt. Pleasant	52	13
Abbey Green	108	71
Whitfield Hall	54	42
Farm Hill	127	70
Penlyne Castle	65	21
Woburn Lawn	81	20
Epping Farm	104	50
Mt. Charles	93	nd
Flamstead	22	nd
Chesterfield	nd	nd
Chester Vale	179	60
Clydesdale	114	56
Industry	54	11
Strawberry Hill	83	31
Westphalia	99	31
Orchard	222	65

Source: CO, 137:242, 238–241.

of the workers. On the former development, he reported that the larger estates had more "hands than they can afford to pay at the present rate of wages." On the latter, he reported that on other properties, "great difficulty is found in procuring labour, as the Negroes are certainly more intent in working their own grounds than cultivating the estate, finding it more profitable than any wages that could be offered them" (CO, 137:242, 241).

The magistrates assigned the task of monitoring the transition to a wage labor system were not always welcomed by the planters. In compiling the information concerning the availability of labor reproduced in Table 12, Henry Kent met resistance from several estate mangers. For example, he was denied access to information on Sherwood Forest and Radnor, which were both managed at the time by Andrew Murray. Kent reported that he was denied "any information respecting these properties by Mr. Andrew Murray . . . the information [was] obtained from the Head people, which I believe to be correct." By "Head people" Kent meant the African Jamaican labor supervisors, who were known as drivers under slavery. He faced the same kind of difficulty with Alexander Bizzet, the resident proprietor of Chesterfield, who reportedly "decline[d] to give any information and demand[ed] to know what the Government ha[d] to do with his private affairs" (PP, 137:242, 241).

By the middle of the 1850s, the Jamaican coffee industry was on the verge of collapse. In 1855, Henry Kent, who in 1835 had optimistically opined from Chester Vale that "the negroes in general throughout this district seem well disposed, and I think perform their labor willingly" (PP, 1835:50, 258), bemoaned "[i]t is difficult now to procure labour, the able bodied people having mostly left the estates. . . . The grog shops spreading throughout this Parish . . . will baffle all the efforts of the Philanthropist to improve the moral condition of the people" (CO, 137:327, 97). An unsigned dispatch from 1857 sheds light on how ecological degradation escalated the decline of the coffee industry. "The heavy rains and tempestuous weather experienced since October has [sic] done incalculable injury on the coffee plantations, strewing the ground with the green fruit, and where the fields have been cleared washing away all the rich soil leaving the roots of the trees bare . . . there is great difficulty experienced in carrying on cultivation; the able bodied Negroes having left the estates and idling their time away upon their small freeholds."

The final blow to the Jamaican coffee industry was the abolition by Parliament of preferential duties on colonial produce. In the late 1840s and 1850s, the Jamaican planters could not compete with the cheaper coffee produced with the fresher soils and contract labor of Ceylon and the vast slave-labor production of Brazil. When asked by a Parliamentary committee to account for the rapid abandonment of Jamaican coffee plantations, Alexander Geddes testified that in Jamaica many "of the coffee plantations are worn out. Coffee is planted in what is called virgin soil, and can only be once planted. The great cause for the abandonment of the coffee plantations is the low price now obtained here [London], owing to the introduction of slave and Ceylon coffee" (PP, 1847–1848:23). According to Geddes's testimony, the abandonment had begun immediately following the end of apprenticeship.

By the 1860s, the once-flourishing Jamaican coffee industry had all but disappeared. In the Yallahs region, most of the coffee plantations had ceased production. Of the 38 coffee estates that had been reported in 1836, only 2 (Arntully and Green Valley) were in production in the 1860s. Change had come to the region. In the next section, I will explore in more depth the spatial transitions that occurred during the period between the abolition of slavery and the Morant Bay Rebellion.

PRODUCTION OF LANDSCAPES AND SPATIALITIES

With these historical developments in the Yallahs region sketched out for the postemancipation period, the question now turns to a consideration of how space and spatial manipulations were affected during this period of social renegotiation. As was the case in the preemancipation era, spatialities were differentially experienced by African Jamaican and European people. By virtue of their having been enslaved prior to 1834, the Europeans felt that the African Jamaican population should remain a laboring class, while they themselves, the Europeans, should remain the supervising, dominant class. The redefinition of space in the years following emancipation reflected the contradiction between the European desire to create a dependent working class, and the African Jamaican conception of what defined freedom.

In examining how the productive spaces on Yallahs region coffee plantations were transformed in the period 1834–1865, I will consider data from the same data sets used to discuss space in the preemancipation period, as discussed in Chapter 6. In doing so, I will use the cartographic, documentary, and material records of several Yallahs estates to provide an interpretation of the relationship between social and spatial restructuring in the postemancipation period.

By the time that slavery was abolished, most of the Yallahs coffee plantations were mature estates. With emancipation, these estates were transformed into wage-paying agricultural concerns. In part because of a more direct relationship between capital paid for labor power and production, many of the coffee planters had their properties surveyed. In contrast to the few estate plans that exist for the preemancipation period, detailed plans depicting various spatial elements exist for several Yallahs plantations, including Sherwood Forest, Sheldon, Arntully, Green Valley, and Chesterfield plantations. In contrast to the preemancipation plans and plats, these estate plans do not always focus on the boundaries or even the extent of the various estates, but provide much more detail on the internal layout of the plantations, particularly the shape, dimensions, and size of the coffee fields. This may reflect a change in logic of estate ownership, as more importance was placed on the efficiency of production rather than the quantity of land owned by an estate proprietor. Interestingly, there is a significant temporal gap in the cartographic record. Most of the surviving postemancipation estate plans date from the 1850s or later, suggesting that in the period of most rapid economic decline on the coffee plantations, much more precise control over the material space of the coffee estates was attempted.

The Material Space of Coffee Production: Coffee Fields

One of the crucial elements in the construction of wage labor is the detailed quantification of production. This is reflected in the post-emancipation estate maps of the Yallahs region, which place a great importance on the exact quantification of the amounts of land contained in the various coffee fields. By exerting this type of control over space, the plantation managers could, by extension, exert control over labor, by quantifying the amount of land harvested. Such quantifications were used in determining the scales of wages to be paid to the emancipated laboring class.

No plans of Yallahs estates depicting coffee fields and dating to the apprenticeship period have yet been identified. The earliest post-emancipation plan so far discovered is a depiction of Arntully dated January 1839 (STT 7A). This rendition of the estate by an unknown surveyor depicts several cartographic features, including 15 coffee fields, comprising 86¾ acres of Arntully's total of 619 at that time. The coffee fields were moderately sized, ranging from 2¼ to 10¼ acres. Although somewhat a matter of conjecture, since this plan was drawn a mere 5 months after the end of the apprenticeship system, it may well indicate how space was divided into coffee fields during this period of transitional labor.

A rendition of Chesterfield drawn by James Smith (apprentice to John M. Smith; Higman, 1988:60) in 1854 (Figure 37) exhibits the developing preoccupation with precise spatial control over coffee fields. According to the plan, at this point in time, the estate belonged to Charles Barclay. Although the boundaries of the estate are depicted, nowhere on the map is the entire acreage of the estate given. Each of the estate's coffee fields is carefully depicted and quantified. All told, 17 distinct coffee fields containing 66 acres, 2 roods, and 24 perches are depicted (1 rood=¼ acre; 1 perch=30¼ square yards). Several of the larger fields are shown subdivided. For example, Jackson Field is divided into five coffee pieces, ranging in size from 25 perches (approximately 756 square yards) to just over 1 acre. Similarly, Coppard Field is subdivided into three coffee pieces, each containing just over 2 acres. According to this depiction, the average size of a coffee field on Chesterfield in 1854 was just over 4 acres.

A similar plan of Abbey Green rendered in 1852 also by James Smith for his father, the surveyor John M. Smith (b. 1799, d. 1854; Higman, 1988:297), demonstrates a similar set of spatial phenomena (Figure 38). According to the plan's legend, the estate was managed at this time by Henry Coppard. As was the case with the Chesterfield plan,

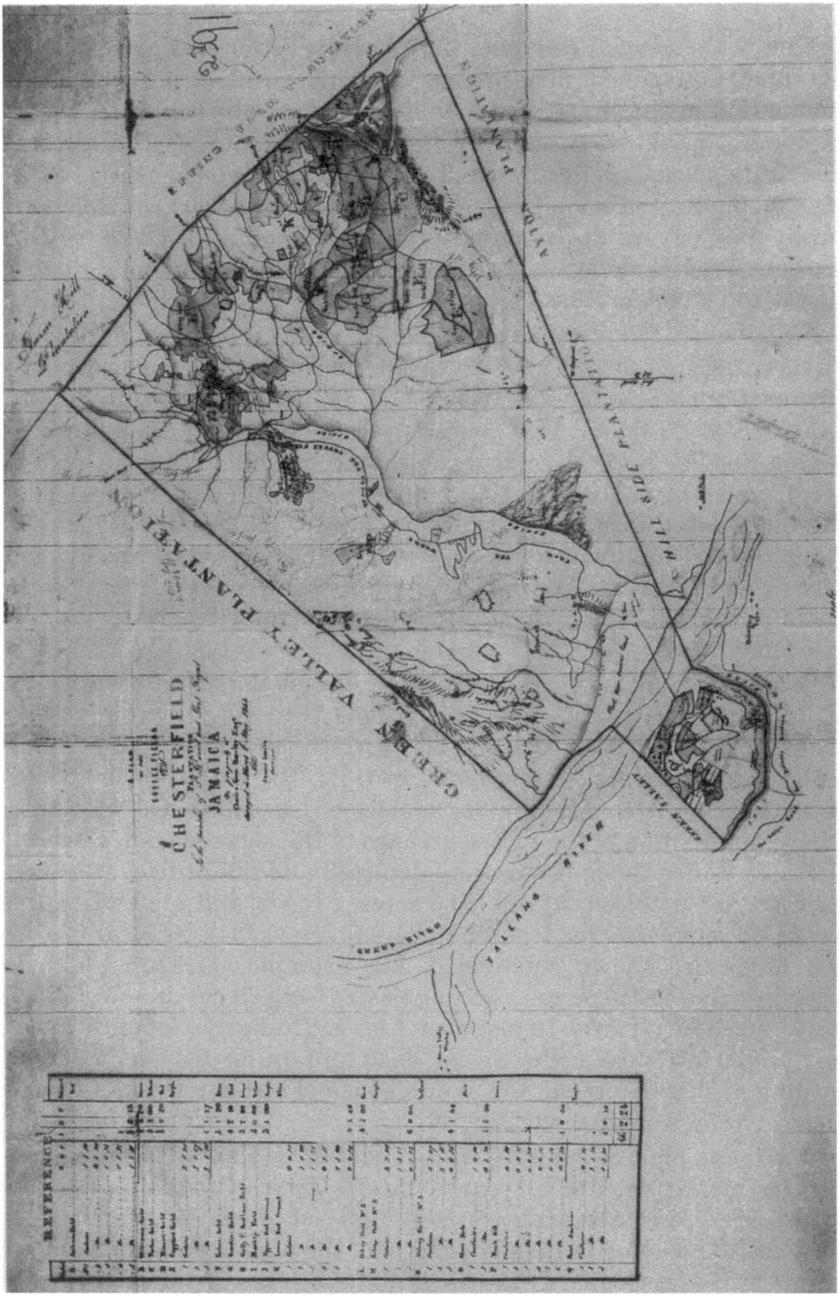

Figure 37. Chesterfield estate plan by James Smith, 1854. Courtesy of the National Library of Jamaica.

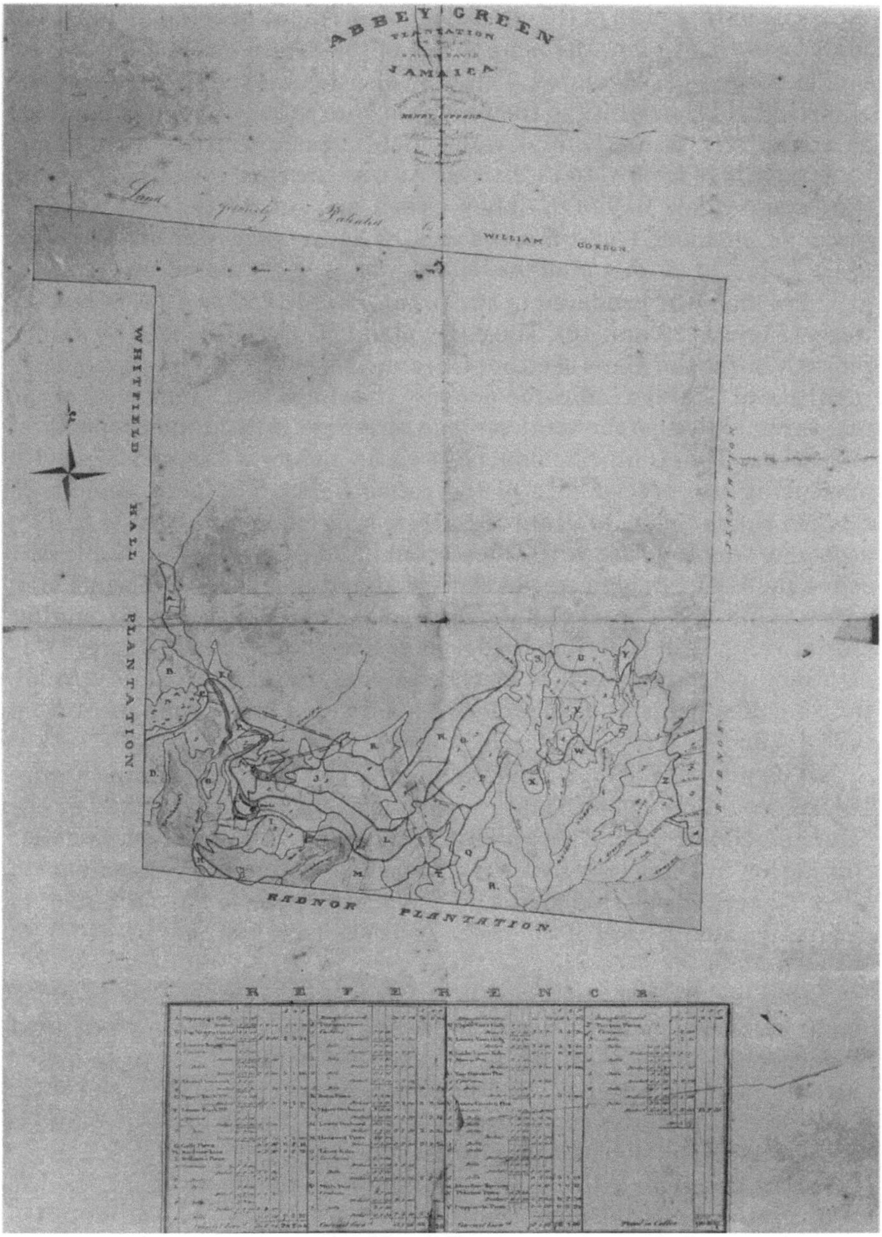

Figure 38. Abbey Green estate plan by James Smith, 1852. Courtesy of the National Library of Jamaica.

the focus of this plan is the precise measurement of coffee fields. Like the Chesterfield plan, the boundaries of the estate are laid out but no total acreage was calculated. This plan also depicts the size and acreage of coffee fields, accurate to the perch. In sum, Abbey Green is depicted as containing 26 individual coffee fields, each with its own name, ranging in size from ½ to 13½ acres. As was the case with Chesterfield, the larger coffee fields at Abbey Green are subdivided into smaller pieces of between 1 and 5 acres each. The average size of the coffee fields depicted on this plan measured just over 3½ acres.

Sheldon was rendered in two plans, dated 1852 and 1856, respectively (Figures 39 and 40). The early plan was also rendered by James Smith. Unlike the plans of either Chesterfield or Abbey Green, the 1852 rendition of Sheldon indicated not only the shape and boundaries of the plantation, but also the total acreage contained within the property — 509 acres. The 1852 Sheldon plan again exhibits the concern with measuring the precise size of the coffee fields. The plan depicts 25 distinct coffee fields, totaling 161 acres, ranging in size from ¼ to 32¾ acres. As was the case with Chesterfield and Abbey Green, the larger coffee fields at Sheldon were subdivided and measured. Nelson Field, which contained a total of 32¾ acres, was subdivided into 20 smaller pieces, ranging in size from ½ to 3 acres. Newland Field (22 acres) was also subdivided into smaller pieces, ranging from 1 to 2¾ acres. In all, the 25 coffee fields at Sheldon were subdivided into 61 pieces of from ¼ to 6¼ acres.

The plan of Sheldon rendered in 1856 by R. J. Robertson (Figure 40) suggests that a significant proportion of the coffee fields had either been abandoned or sold. According to this plan, of the 161 acres under cultivation in 1852, only 60¼ acres were still defined as coffee fields in 1856. Nevertheless, this plan also depicts the precise measurement of productive space; each of the coffee pieces depicted ranges in size from ¾ to 2½ acres.

Similar divisions of space in the 1850s are represented in plans rendered by the younger Smith for Ayton in two plans drawn in 1851 (Figure 41) and for Windsor, an estate that bounded Arntully in the extreme eastern section of the study area. As was the case for the other plans drawn by Smith in the 1850s, the focus of each of these plans is the detailed depiction of coffee fields.

Two detailed Yallahs estate plans are known to exist for the study period, both dating to the 1860s. An exceptionally well-detailed plan of Arntully was drawn by Felix Harrison in 1860 (Figure 42). This plan suggests that the division pattern for coffee fields exhibited by Smith's 1850s plans continued into the 1860s. The plan represents

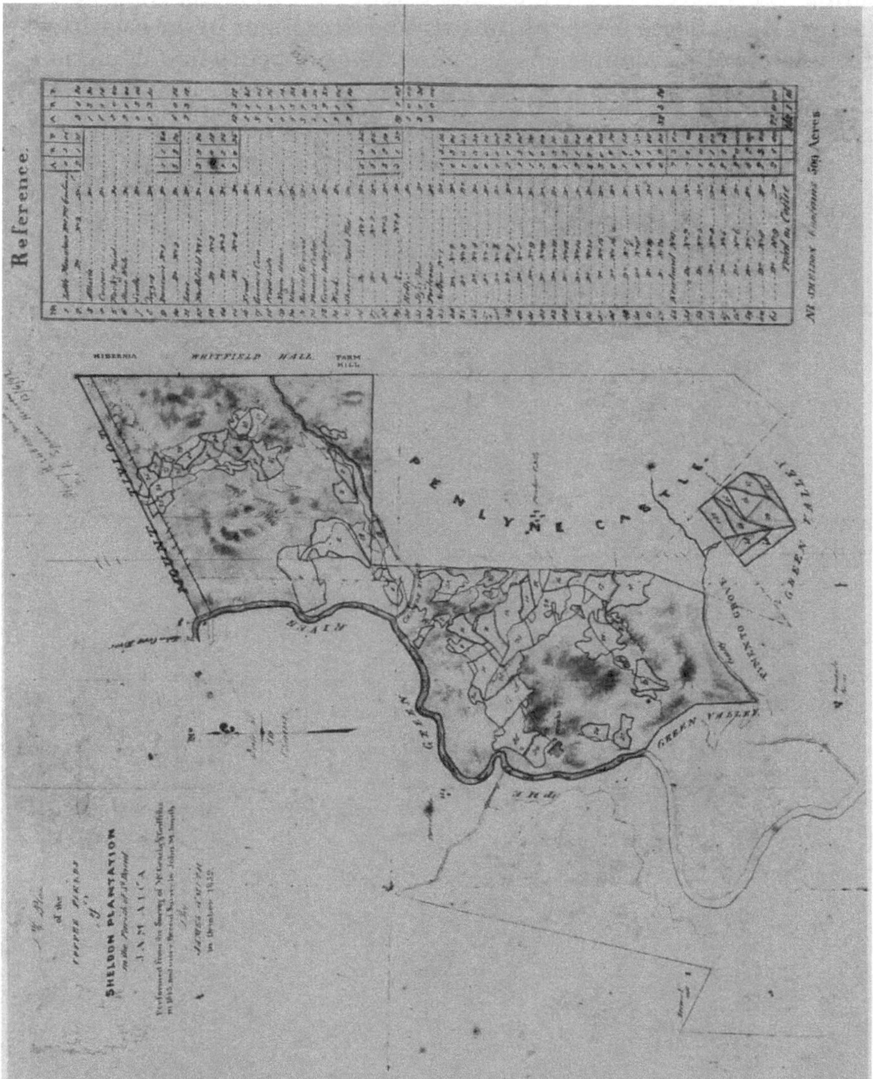

Figure 39. Sheldon estate plan by James Smith, 1852. Courtesy of the National Library of Jamaica.

a division of space on the plantation that includes 29 coffee fields, totaling 111¾ acres, subdivided into a total of 70 pieces, ranging in size from ¼ to 4 acres. Perhaps the extreme expression of the cognitive definition of coffee plantations as coffee fields appears in a depiction

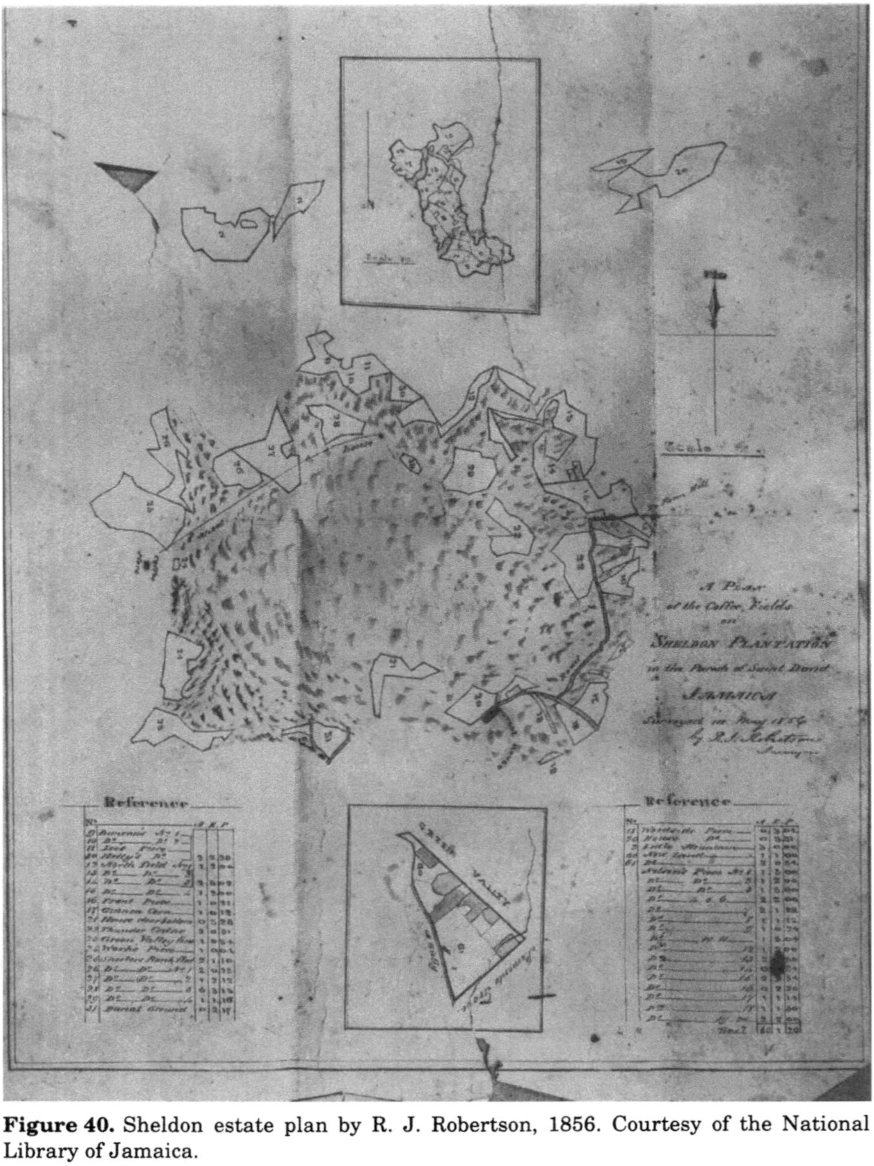

Figure 40. Sheldon estate plan by R. J. Robertson, 1856. Courtesy of the National Library of Jamaica.

of New Battle, an estate bounding Radnor to the south. The only landscape features depicted in this plan are coffee fields and the coffee works. Twelve fields, containing 42¼ acres and ranging from ½ to 11¾ acres, are depicted. As no estate boundaries or topographic

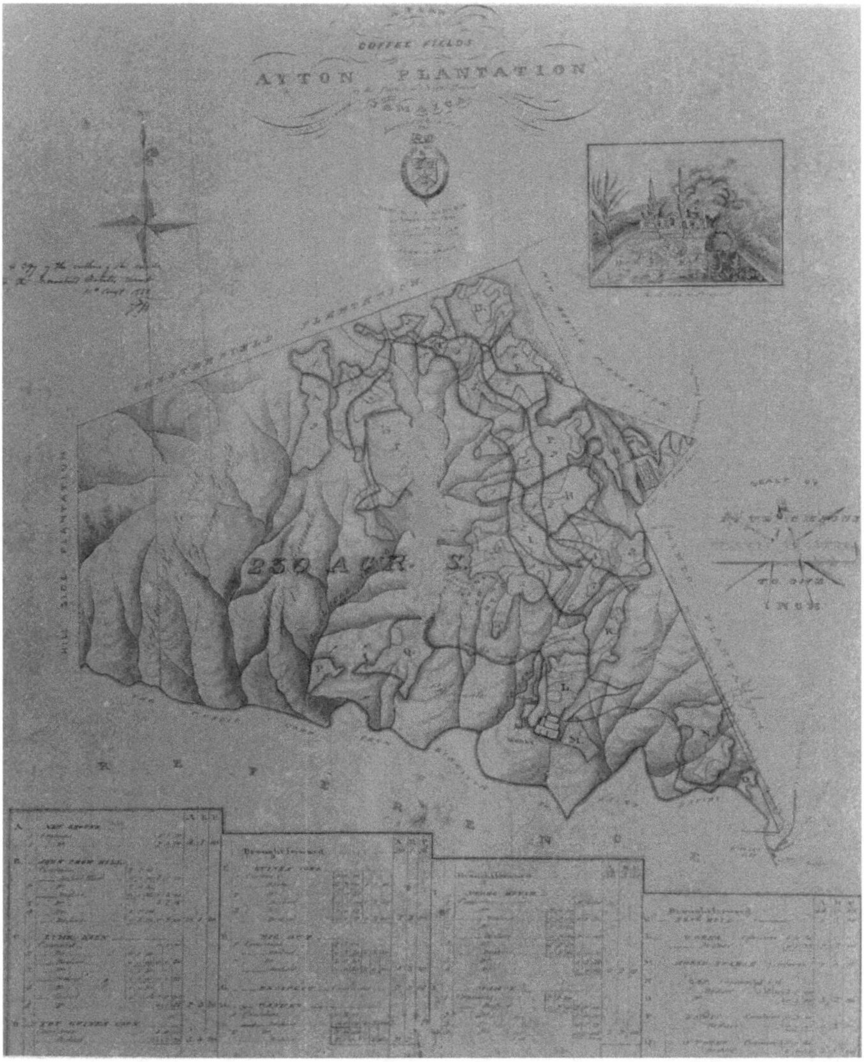

Figure 41. Ayton estate plan by James Smith, 1851. Courtesy of the National Library of Jamaica.

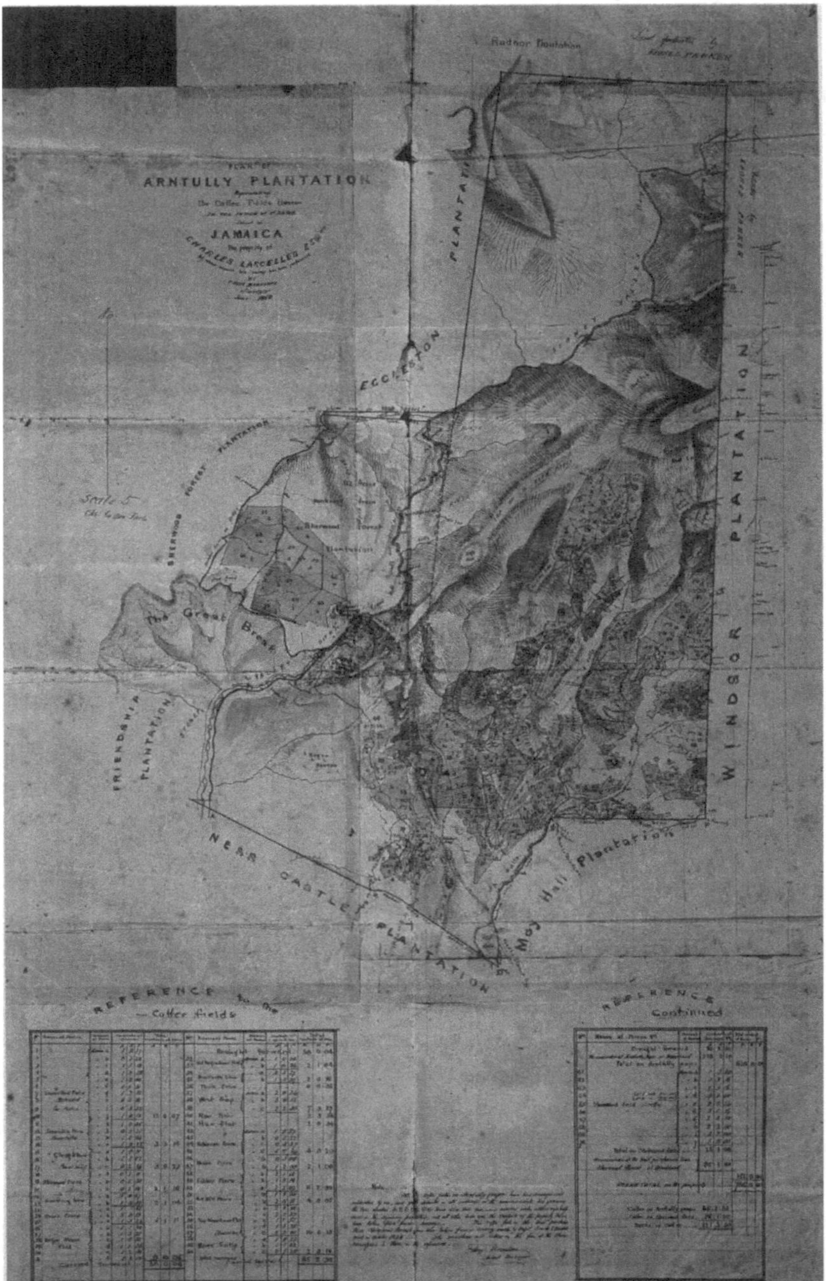

Figure 42. Arntully estate plan by Felix Harrison, 1860. Courtesy of the National Library of Jamaica.

features are depicted, this plan leaves one with the impression that the coffee fields existed in spatial isolation, some connected to nothing in space, not even other coffee fields.

When compared to the preemancipation estate plans discussed in Chapter 6, the estate plans rendered in the decades following emancipation demonstrate an increased importance in defining coffee fields on estates. Many of the postemancipation plans privilege the depiction of coffee fields over all other landscape features. Several of the plans depict nothing but the shape and size of coffee fields. In the absence of slave labor, the quantification of the spaces of production became increasingly important to the landowners. As the use of space for coffee production by wage laborers became increasingly difficult for the planters, their precise control over that space increased in importance.

The Material Space of Coffee Production: Coffee Works

In all likelihood, the plantations that remained in operation throughout the postemancipation period used the coffee works that were constructed under the slave regime. The documentary record, as surveyed by Montieth (1991) and Higman (1988), seems exceptionally silent on the issue of coffee works. The one area of change recognized by these historians involves the speed of coffee processing (Montieth, 1991:188). Although Montieth suggests that technological innovation may have decreased the number of laborers required to operate the machinery, she gives no indication of change in the spatial organization of coffee works.

My archaeological survey of the three plantations works also found no evidence of substantial change in the spatial organization of plantation works. There are no signs of alteration, such as an increased number of pulping mills and water courses, or an increased number of barbecues on plantations. In the Yallahs region, the spatial order of coffee plantation work remained essentially that designed and executed for enslaved labor.

The Material Space of the Elites

Unlike the coffee works, the domestic material spaces of the elites do reflect the change in labor regime. Of the different buildings constructed on the plantations for use by the elites, overseers' houses seem to have been the most durable. No great houses constructed during the period 1834–1865 are known in the Yallahs region; however, it seems likely that some of the buildings constructed prior to emancipation were still occupied during this time.

The architectural record of the overseer's house at Sherwood Forest demonstrates that this class of structure changed over time on plantations that continued production well after emancipation. The original overseer's house was constructed of fieldstone covered with lime plaster; the exterior walls of this building were 21" thick. The small building contained two rooms. The southern room was the domestic space for the overseer. A veranda provided the overseer with a southern view that encompassed the coffee drying platforms, or barbecues. An interior wall 18" thick separated the overseer's room from the pulping mill, which obviously was contained in the same building as the overseer's domestic space. The original great house on this plantation was located approximately ½ mile away from this complex well out of view of the industrial complex. As long as he was in his room or on his veranda, the overseer on this plantation could survey the production of coffee. From his veranda he could monitor the coffee crop as it dried. From his room, he could watch the coffee as it was pulped and channeled to the drying platforms. This spatial arrangement created an intensive capacity for the surveillance of production.

The second phase of construction at the overseer's house was much more gracile, and served to remove the overseer from the direct surveillance of production. The second house was literally built on top of the first; the older house serves as part of the foundation for the newer structure. This later building is of timber construction, and itself experienced several phases of construction. The pulping and grinding mills remained in the building, but were now located on a lower floor, separating the domestic space somewhat from the direct activity of production. Interestingly, the veranda of the second building is oriented to the east. The viewscape from this perspective encompasses a scenic view of the valley and surrounding hills; it takes a conscious effort to supervise production from this vantage point.

This new spatial arrangement reflects a change in the relations of production that developed after emancipation. Beginning in the late 1840s, Sherwood Forest began acquiring the estates that bounded it to the east and south; the previously independent estates Arntully and Eccleston were defined in the crop accounts of the 1850s as appendages to Sherwood Forest. The comfortable wooden structure may have developed into a postemancipation great house from which the affairs of the several plantations were managed. In this scenario, the direct surveillance of wage-laborers may not have been of primary concern. In his discussion of Arntully, Higman (1988:175) notes that no great house was indicated in the 1839 plan; this may substantiate the hypothesis that during the 1840s, at least, the affairs of Arntully may have been directed from this structure.

The Material Space of the Enslaved

Following the end of slavery, Europeans began to take much greater notice of the uses of space made by the African Jamaican population. In large part because of the interest on the part of the Europeans to transform the spaces used by the laboring population into commodities for the redistribution of wealth through rent and taxes from the workers to the planters, it is possible to say more about the material spaces of African Jamaicans in the postemancipation period than it is during the time of slavery.

Idealized Domestic Space of the Enslaved

As was the case with the preemancipation materials, it is difficult to derive many conclusions about the organization of domestic space within the houses of the African Jamaican population of the Yallahs region without the benefit of excavation. The plans of Ayton and Abbey Green rendered by James Smith in the 1850s indicate the location of "Negro Houses"; in each case, the houses appear clustered together, much as they had in the preemancipation villages. As these representations lack detail, it is difficult to determine whether these plans accurately represent the number and location of individual houses, or whether they are merely conventional depictions of structures indicating the general location of the quarters. The Abbey Green plan (Figure 38) shows the relative location of 17 individual buildings. In January 1839, Henry Kent reported that there were 100 formerly enslaved people residing on the estate; 8 people had by that time chosen to leave the estate.

Assuming the accuracy of these house counts and combining them with Kent's figures, there were 5.9 people per house on Abbey Green in 1852. In considering the two plans at Ayton, this formula results in totals of 2.1 and 3.5 people per house, respectively. These numbers are consistent with a variety of possible household membership structures, including both nuclear family and extended family structures. Thus, the maps may represent the number of houses and hence the household size in the postemancipation Yallahs region. Excavation would be a way, in the future, to check the accuracy of these maps and detail the material conditions of domestic life on the estates.

Several data sets can be used to interpret the material nature of the provision grounds used by the African Jamaican population in the Yallahs region following 1834. One of the most significant is an estate plan of Green Valley rendered in 1837, during the period of the apprenticeship (Figure 43). This plan was drawn specifically to measure the extent of the provision grounds cultivated by the laborers attached to the estate. According to this plan, the only rendition of provision grounds known to exist for the Yallahs region, the section of the estate utilized for provisions was divided into 18 pieces, ranging in size from 3½ to 13½ acres. This document is all the more significant in as much as it identifies the types of foodstuffs produced on each parcel and identifies the people who worked each piece of land. The heading above the lists of names reads "[t]he following are the names of those apprentices who possess grounds . . ." indicating that from the perspective of the planters, the laborers had some claim to possession, if not ownership, of their provision grounds. These data are shown in Table 13.

Table 13. Provisions Grown, Size of Plot, and Number of Cultivators on Provision Grounds at Green Valley, 1837

Plot	Provisions grown	Acreage (acres:roods:perches)	Number of cultivators	Acres per person
1	Plantains	4:1:06	9	0.48
2	Pidgeon and congo peas, ruinate, grass	7:2:38	9	0.84
3	Congo peas, plantains	12:3:33	17	0.75
4	Yams, cassava, corn, canes, plantains	12:2:52	16	0.79
5	Yams, cassava, corn, canes, plantains	13:2:00	35	0.34
6	Cocoas, yams, ruinate, grass	16:0:00	8	2
7	Plantains, cassava, cocoas, yams, canes	9:0:06	20	0.45
8	Plantains, cassava, cocoas, yams, canes	12:0:11	17	0.71
9	Plantains, cassava, cocoas, yams, canes	9:0:00	9	1
10	Plantains, grass, ruinate, wood	9:2:10	19	0.5
11	Yams, cassava, canes, congo peas, fruit	8:2:14	19	0.45
12	Yams, cassava, canes, congo peas, fruit	3:2:00	8	0.44
13	Peas, yams, ruinate	10:2:07	12	0.88
14	Peas, yams, ruinate	7:2:10	7	1.1
15	Peas, yams, ruinate, sweet potatoes, bananas	13:1:22	14	0.95
16	Peas, yams, ruinate, sweet potatoes, bananas	9:0:00	6	1.5
17	Ruinate, grass, cocoas, plantains	6:1:29	7	0.9
18	Grass, ruinate	6:0:20	1	6

Source: STA 366.

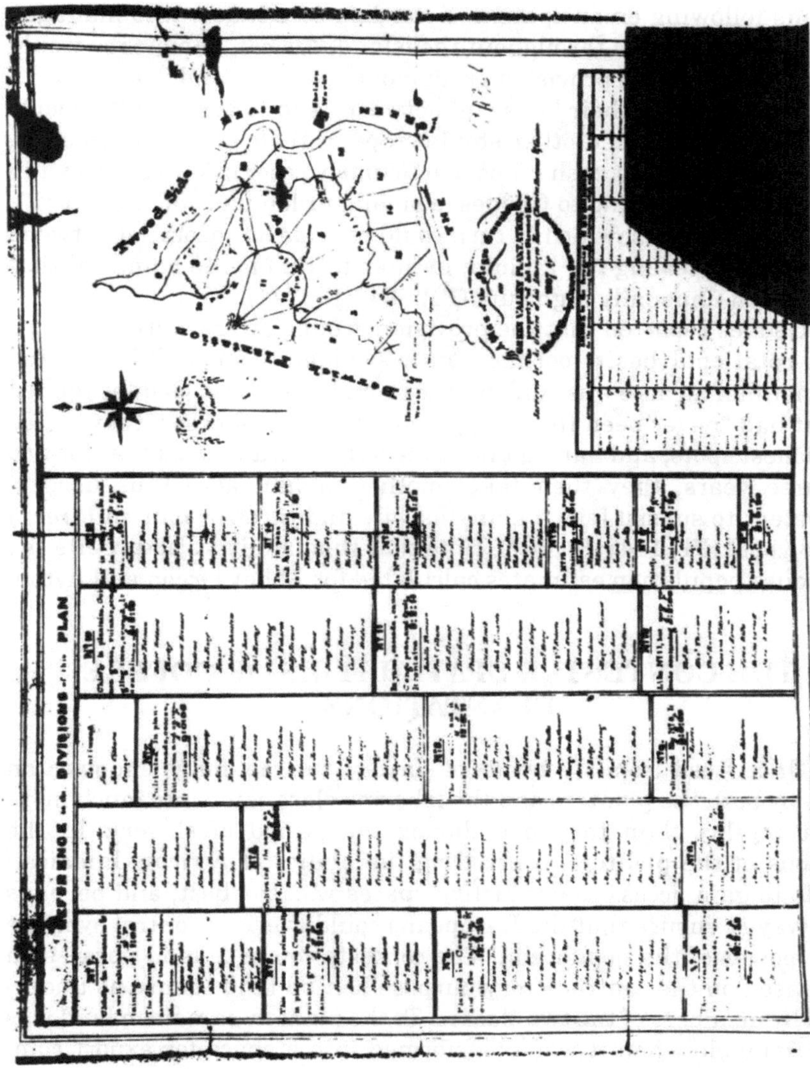

Figure 43. Green Valley estate plan by Edward Geachy, 1837. Courtesy of the National Library of Jamaica.

The testimony given by Alexander Geddes to the Parliamentary committees on Jamaican cultivation provides some insight as to the extent and nature of provision cultivation that occurred in such grounds following emancipation. According to Geddes, who managed several coffee estates throughout the island, African Jamaicans would generally cultivate a provision ground of between ½ and 2 acres and a garden attached to their house, of between ⅛ to ¼ acre. While people would grow tubers and other staple crops in their provision grounds, Geddes testified that cash crops, particularly coconuts, were grown in the gardens. According to Geddes, a small garden plot could support a dozen coconut trees, which in turn could provide an income of between £10 and £30 sterling per annum; in later testimony he admitted that this was an unusually high sum (PP, 1842:13, 741).

Geddes reported that the provision grounds were cultivated on a short rotational basis; perhaps on a swidden system. When asked whether half an acre was sufficient to provide a laborer, Geddes replied that it had "never been the custom to restrict the negroes; they culled the richest spots, and having taken one half acre, and used that for two or three years, they then take another; a half acre thus used is competent to subsist [sic] a man, his wife and three or four children a whole year" (PP, 1842:13, 467). Note that this family structure fits the per-house population estimates calculated for Abbey Green and Ayton.

THE CONTESTED SPATIALITIES OF COFFEE PLANTATIONS

Once freed from the bonds of slavery, the meanings endowed on space by both the planters and the laboring class came into much more direct conflict than they had during the period of slavery. In the opinions of the planters, controlling the ability of African Jamaican people to gain access to productive spaces was the best, and perhaps only, way to ensure that the Europeans could exercise control over the labor power of the population. The contestation over land possession was based on the contradictory purpose behind land tenure as understood by members of the two classes. To the planter, Jamaican land was most valuable when it was transformed into a space for export commodity production. To the laborer, access to land created a spatiality of independence. If a person or family could control enough land to produce enough food for subsistence, and provide surplus for exchange in local markets, that person or family could become self-sufficient. Much of the history of Jamaica since 1838 has been driven by this contradiction. In the mid-nineteenth century, this contest was evident in conflicts over the negotiation of spatial control over land through purchase by the laborers, rent charged to people occupying productive spaces, and the assessment of land taxes.

Creating Self-Sufficiency through Freeholds

From the inception of the apprenticeship system, it was the desire of many African Jamaican laborers to create a tangible, material independence from their former slave masters. This was expressed by some through the purchase of their own freedom during the apprenticeship, as discussed above. Once free, many of the formerly enslaved people recognized that the way to gain true independence from the planters was to gain legal control over land. Holt argues that this process began even before full emancipation, as some laborers planned ahead by saving money in order to purchase lands after their apprenticeship ended (Holt, 1992:143). As early as 1843, the planter class recognized that the legal acquisition of land by the peasantry threatened the hegemony of the planters. In a pamphlet published in 1843, Robert Paterson, who described himself as a "resident proprietor," stated that the "chief object which has induced the negro to labor, since his freedom, has been to procure the means of purchasing from one to five acres of land, that he might there erect a hut, evade the payment of rent, and sit down in imaginary independence" (Paterson, 1843:4).

Paterson describes the nature of the contradiction between capital and labor as it was experienced quite succinctly: "It delights the eye of the traveler to see patches of land cleared, houses erected, and their occupants living in seeming contentment; but alas! these appearances are not solid — the staple productions of the soil are neglected — and unless this evil be immediately remedied, the country . . . will ultimately become a colony of small settlers, without capital, without enterprise, without circulation of any kind to stimulate them to improvement" (Paterson, 1843:4). Paterson's observations describe the basic contradiction in land tenure. While the elites saw land as a means for producing commodities and thus stimulating the circulation of capital, the African Jamaican settlers saw land as a means of creating self-sufficient farms, independent from the Eurocentric political economy.

Alexander Geddes also commented on the availability of cheap land and the threat that it imposed to the hegemony of the planters. In his 1842 testimony before Parliament, when asked what might happen if the estates lowered wages throughout the island, Geddes stated that the people would "betake . . . to the cultivation of waste lands, which can be purchased at nominal prices. One great source of our misfortune in Jamaica is, that the negroes were taught to place their sole reliance for subsistence upon the cultivation of the roots common to the country; and it is impossible to eradicate that habit; it is now our great bane" (PP, 1842:13, 469). Geddes's comments reflect an ironic development. The Europeans had believed it cheaper and more expedient to require the working population to produce its own food under slavery. This created a sense of spatial independence in the provision grounds; once emancipated, this definition of social space had not only been retained by the population, but it had become the focus of the quest for freedom.

The contradiction in interpreting the meanings of social space in the provision grounds was anticipated among the planter class. For example, under an examination in 1835 on the working of the apprenticeship system before a Parliamentary committee, Dr. John Pine, who had provided service in both St. David and St. Thomas-in-the-East, stated that the apprentices "work too much land. In my neighborhood, it is too profitable an employment; as long as they have indiscriminate access to any quantity of good land, they can never be expected to work generally for wages" (PP, 1835:50, 94). In a proclamation to the apprenticed laborers printed in the *Morning Journal*, a Jamaican newspaper, on July 30, 1838, the governor, Sir Lionel Smith, expressed this concern. In discussing the postemancipation circumstances of the population, Smith implored the African Jamaican people to remain on the estates "on which you have been born, and where your parents are buried." He warned the laborers, however, to recognize the hegemonic control over space exercised by the proprietors: "you must not mistake, in supposing that your present houses, gardens, or provision grounds, are your own property. They belong to the proprietors of the estates, and you will have to pay rent for them in money or labour, according as you and your employers may agree together" (*Morning Journal*, July 30, 1838).

In this context, assessing rents can be interpreted as a means by which the planters attempted to mitigate the contradictions between the social spaces understood by the laborers and the social spaces they themselves understood. In the years following the end of the apprenticeship period, many proprietors of the Yallahs estates attempted to use this means to continue to exert control over space, and thus control over "free" labor. In reports filed to the Colonial Office in 1839, Henry Kent discussed the various strategies used on the Yallahs estates in both St. David and Port Royal. According to his report, the most common way rent was collected in St. David was by compelling laborers to work 1 day per week in return for the use of the houses and grounds that the laborers occupied. This system was in place on Sherwood Forest, Minto, Whitfield Hall, Somerset, Penlyne Castle, Woburn Lawn, Epping Farm, and Old England. No rents were assessed by Mt. Teviot, Radnor, Mt. Pleasant, Abbey Green, or Farm Hill (CO, 137:242,

240–41). This system was also common in Port Royal, although the manager of Mt. Charles, George Wright, assessed a weekly cash rent of 1s 8d per family for the use of a house and grounds, an additional 1s 8d to pasture a horse and 1s 3d to pasture a donkey. The manager at Clydesdale charged each able-bodied worker, excepting children, 1s 8d per week. The managers and proprietors of Flamstead, Chesterfield, Chester Vale, Industry, Hall's Delight, Strawberry Hill, Westphalia, Orchard, Mahogoney Vale, and Windsor Lodge all assessed 1 day's labor per week, while no rent was assessed at either Pleasant Hill or Resource (CO, 137:242, 238–39).

Kent's reports also discuss how well the mitigation of the contradictions in social space were working. For example, on Mt. Charles, Kent reported "great discontent on this property in consequence of Mr. George Wright, the attorney, making the heavy charges [for rent] against the people and which the overseer Mr. Anderson stated in court was for their making so many complaints to the magistrates" (CO, 137:242, 241). Both Wright and Anderson, on Anderson's own admission, apparently hoped to teach the working class a lesson by raising their rents as a consequence for their attempting to exercise the civil right they had been granted under the Emancipation Act, to seek redress over labor grievances from the special magistrates. There may have been a similar strategy employed at Flamstead; although no motive is expressed, Kent reported "great excitement on this property in consequence of the Receiver, Mr. Fyfe, doubling their rent." On Pleasant Hill, rent was also used as a means to attempt to coerce the population to conform to the planter's labor demands. Kent reported "their daily task used to be hoeing from 5000 to 4000 pc and as they will not do it since the death of their master Mr. Simon Taylor, the present proprietor Mr. George Taylor is raising the rent."

Several of the proprietors used even more coercive spatial negotiations in the attempt to control labor. For example, Kent reported that with his arrival at Green Valley, the proprietor James Law Stewart had "served [the people] with notice to give up their provision grounds in 3 months from 7 January which has produced great consternation amoung the negroes." According to Kent's account, the manager of Minto utilized a similar coercive strategy: "a great many people [have] left and been ejected — and are denied the right to take out their provisions — I regret that upon the opinion of the Attorney General I can render them no assistance" (CO, 137:242, 241).

Although no explicit motives are given for these actions on the part of the planters, they can be interpreted as attempts to exert greater control over the spaces utilized by the laboring class. Such efforts may have been enacted in the hope of compelling African

Jamaicans to provide more labor on better terms for the estates. Such motivation may be revealed in Kent's comment about Mt. Faraway. The management of that estate was experiencing great difficulty controlling the labor force, as management could not "get them to work on the term offered; [they] will only work to pay their rent and then go off to other properties" (CO, 137:242, 241).

The negotiation of the conditions of labor may also indicate strategies used to mediate the contradictions in the usage and definition of social space. The only clear evidence of such a negotiation for the Yallahs region, however, is for Arntully. In February 1839, George Willis, one of the special magistrates assigned to St. David, reported that management and labor on the estate were engaged in a negotiation over wages. According to this report, the people had worked only 1 week between August 1838 and February 1839, and were currently engaged in a strike against the estate for better wages. Willis stated that the laborers were demanding 1s 8d, approximately 1 day's wages, for the weeding of 100 trees. On nearby Woburn Lawn, the laborers were paid 1s 8d for weeding 125 trees. Unfortunately, there is no indication as to what the managers of Arntully were demanding, or how the dispute ended (CO, 137:241, 29).

The difficulties experienced by the laborers concerning rents may have further motivated many of the people to abandon the plantations in favor of new houses and grounds that they could operate independently of the whims of the planter class. According to Geddes, the rent system may have backfired for the planters, driving more laborers into purchasing their own lands. This is revealed in the following exchange between the Select Committee on West India colonies and Geddes, in 1842 (PP, 1842:13):

> Question: From what you have said as to the dislike of the labourers to pay rent, it appears that it is not your opinion that the system of paying rents has at all tended to inspire them with feelings of local attachment and unwillingness to leave their localities?
>
> Geddes: I consider that the exaction of rent, which was quite unavoidable under the circumstances, has been the means of inducing them to purchase lands.
>
> Question: Rather than pay rent?
>
> Geddes: Rather than pay rent.

The dialectical contestation between the spatialities experienced by the planters and those experienced by the workers, mediated by rent payments and other attempts at spatial control, resulted in a synthesis by which an increasing number of African Jamaican people abandoned estate production in favor of creating their own farms. Although many people probably became squatters in remote areas, few records of this strategy of spatial resistance remain. One strategy that can be considered, however, is the legal purchase of land. The transference of land to the emerging peasant class became a matter of concern for the Colonial Office, as indicated in a dispatch by Governor Charles Metcalfe to the Colonial Office, in which he states that between August 1838 and June 1840, some 2074 small lots of less than 20 acres had been conveyed in Jamaica. Of these new landowners, at least 934 had purchased enough to register as electors, i.e., were enfranchised to vote. Metcalfe remarked that prior to these conveyances, the number of electors throughout the island was 2199, "to which even 934 would be a large relative addition." Metcalfe did not believe that by 1840 the newly enfranchised voters had yet fully exercised their rights. However, the acquisition of legal titles to land was providing an additional benefit of the control of social space: political power (CO, 137:249).

Many planters in Jamaica, faced with economic decline, began subdividing their estates and selling the parcels to the African Jamaican population. For example, Henry Lowndes, a sugar planter who resided in Jamaica for 27 years, had been the proprietor, trustee, attorney, and/or lessee of eight sugar estates in the central parish of St. Thomas-in-the-Vale. Lowndes reported to the Select Committee on West India Colonies that he himself had sold "a great quantity of land . . . in small lots." Lowndes was able to get £3 sterling per acre, but charged an additional £3 to £4 sterling to cover the expense of surveying and transferring the land. Even at this price, the lots bought by the African Jamaican people from Lowndes were usually from 5 or 6 to 10 acres (PP, 1842:13, 369).

Such land conveyances were common throughout Jamaica, including in the Yallahs region. In a report to the Colonial Office in June 1855, Henry Kent reported that "small settlements" were forming throughout the Blue Mountains "on coffee properties which have been sold and cut up into lots to suit the labouring class, at prices varying from £4 to £6 per acre." In acquiring such lots, the laboring population of the Yallahs district were spatially expressing their independence and freedom from the planters, for as Kent put it, the people were "satisfied to sit down on their acquired freeholds and just cultivate sufficient to supply their own wants, with a little coffee to purchase clothing and necessaries, which in its rude cured state they sell in Kingston for 5 or 6 dollars [sic] per 100 pounds" (CO, 137:327, 97). Although Holt (1992:145) reports that at least two estates, Belle Claire and Middleton, were purchased in their entirety by "peasant proprietors," it seems that the more common strategy employed by African Jamaicans in the Yallahs region was to purchase smaller lots.

Processes of spatial independence in the Yallahs region are best expressed historically by Mt. Charles and Mavis Bank, both of which were subdivided and sold off in the postemancipation period, and have since become villages in the modern parish of St. Andrew. An undated

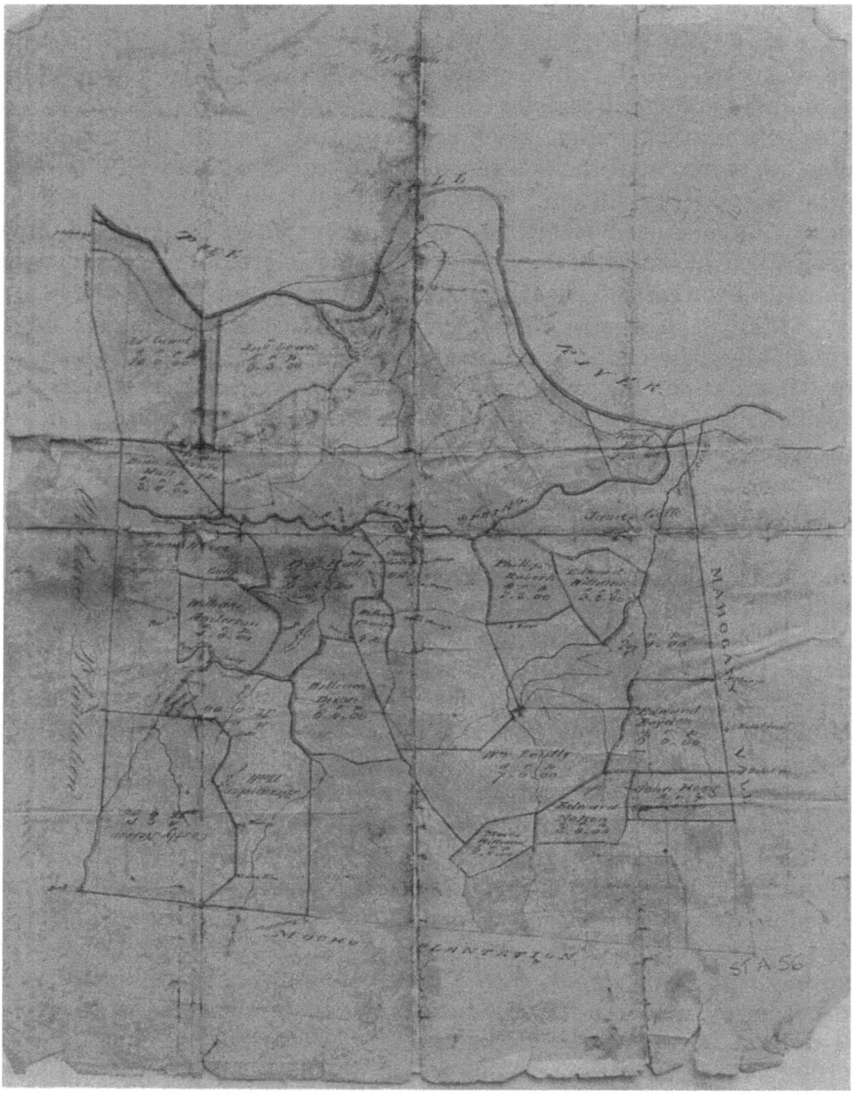

Figure 44. Undated plan of the subdivision of Mt. Charles. Courtesy of the National Library of Jamaica.

plan of Mount Charles (Figure 44) indicates that the estate was subdivided into 19 lots ranging in size from 1 to 20 acres; this plan probably postdates 1838, when the estate was mortgaged. Although the plan is lacking in some detail, it does provide the acreage for each of the subdivided plots. The average plot size was about 5½ acres; each of the purchasers was identified, as shown in Table 14.

The cartographic record of Mavis Bank is more informative on this issue than is that for Mt. Charles. Mavis Bank was first subdivided in 1840 into a few larger plots, including a 60-acre estate purchased by Dr. William Thomson, and 75 acres retained by Robert Sylvester (Figure 45; Table 15). In 1850, Sylvester's parcel was further divided; by this date, the former plantation had been transformed into a village containing approximately 45 homesteads of between 1 and 15 acres (Figure 46).

Although not identified by name in the dispatch by Henry Kent, it is possible that the estate he was referring to in his letter to the Colonial Office was either Mavis Bank or Mt. Charles. If so, the African Jamaicans who purchased lots on these estates were paying between £4 and £6 sterling per acre. The space of these former plantations had been redefined as expressions of spatial independence by a population that had formerly been attached to such estates as slaves.

Table 14. Purchasers of Subdivided Lots on Mt. Charles, c.1840

Name of purchaser	Acres:roods:perches of land
Costly Nelson	13:3:00
Francis Hall	9:1:20
Primus Taite	1:0:00
Diamond Hall	3:0:00
Wm. Anderson	4:0:00
Jane Gilroy	2:1:00
Philip Roberts	2:3:00
James Grant	10:0:00
John Lewis	6:3:00
Edw. Boyden	6:0:00
Edw. Williams	2:3:00
Harry Hall	2:3:00
James Goffe	20:0:00
William Bently	7:0:00
William Dixon	6:0:00
William Thompson	2:0:00
John Hogg	3:0:00
Moses Williams	2:0:00
Edw. Nelson	3:0:00

Source: STA 56.

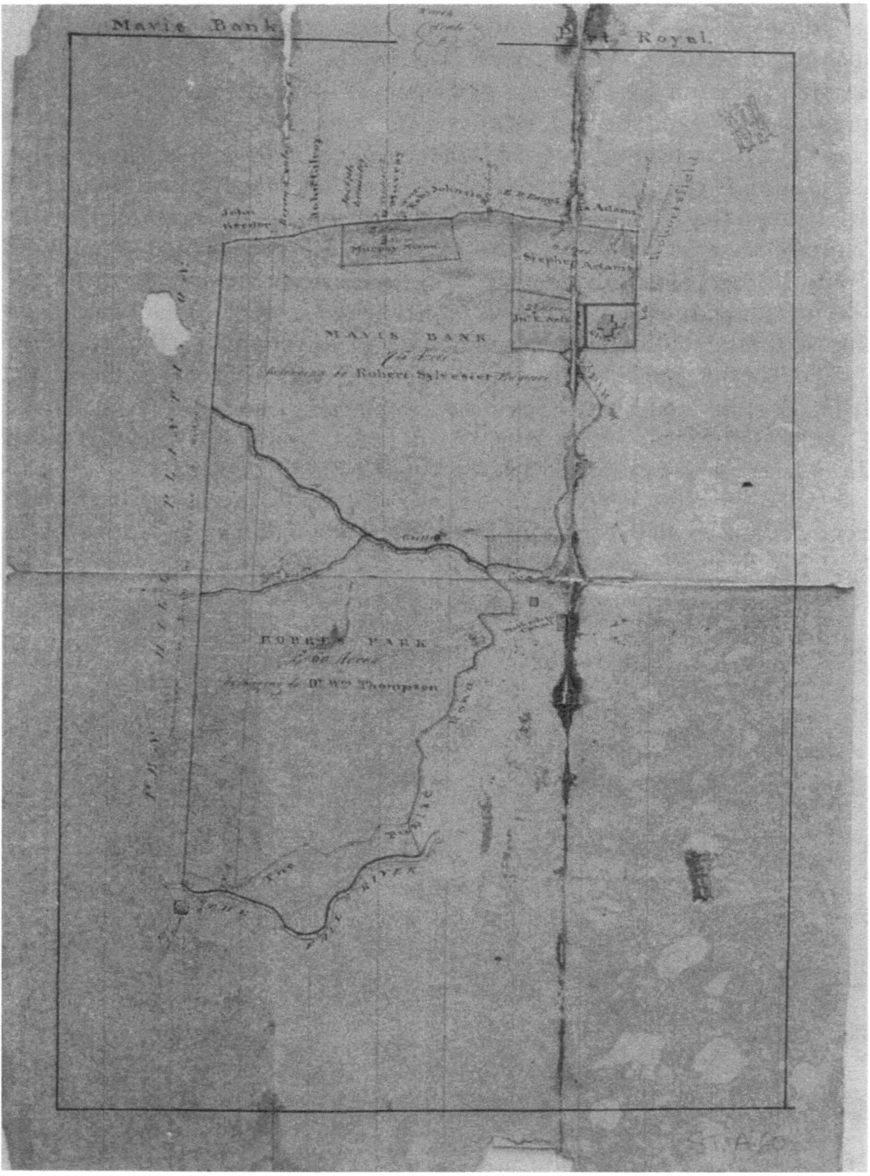

Figure 45. Plan of the 1840 subdivision of Mavis Bank. Courtesy of the National Library of Jamaica.

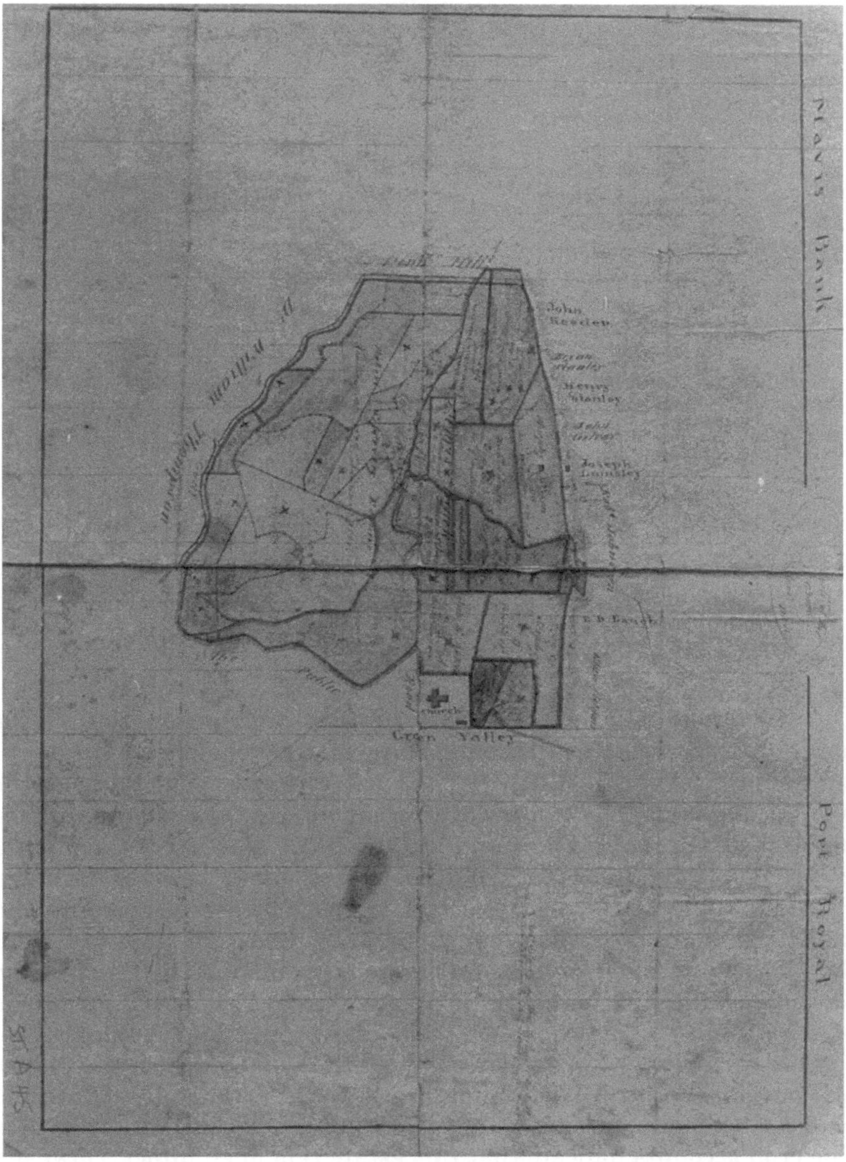

Figure 46. Plan of the 1850 subdivision of Mavis Bank. Courtesy of the National Library of Jamaica.

Table 15. Purchasers of Subdivided Lots on Mavis Bank, c.1842

Name of purchaser	Acres:roods:perches of land
? Walker	10:0:00
Hibbert Nelson	8:2:00
Robert Stewart	5:0:00
? Taite	5:0:00
Moses Williams	6:0:00
John Reeder	7:0:00
Bryan Stanley	5:0:00
Henry Stanley	5:0:00
John Gilroy	6:0:00
Josephy Lumsdew	4:0:00
? Murray	nd
Robert Taite	4:0:00
Peter Hall	nd
Robert Johnston	12:2:00
Eliza Walker	2:0:00
Robert Sylvester	75:0:00
Benjamin Walker	24:0:00
John Williams	0:3:26
Dr. William Thompson	60:0:00

Source: STA 35.

CONCLUSION

The final abolition of slavery in 1834 and termination of the apprenticeship system in 1838 precipitated great changes in the political economy of the Yallahs drainage coffee plantations. Many of the plantations in the region abandoned production in the 1840s and 1850s, not to resume production again until the 1980s. Several of the plantations were subdivided and sold as lots to small peasant farmers. Planters attempted to control access to space by assessing rents and land taxes on the newly developing peasantry. Contradictions in spatial definitions and the uses of space by the laborers and the planters led to conflict that at times escalated into violence. The spatial meanings of some coffee plantations, like Mt. Charles and Mavis Bank, were completely redefined, as coffee estates were subdivided in the 1840s, and developed into village communities. Other estates, like Whitfield Hall for example, pursued a different course following emancipation. The crop accounts indicate that this estate remained intact. However, the record indicates that the formerly enslaved workers and their descendants rented cottages and land from the estate, probably the very houses and provision grounds that were developed under slavery.

Unfortunately, the records do not indicate the size of the rented grounds or their inhabitants. We can surmise, however, that the renting option was not a popular one with the emerging class of farmers. In 1842, the estate collected £36 in rent. The total amount of rent collected declined each year through 1846, when a mere £8.2.0 was collected. No record of rent collected appears in the accounts following this date, which may indicate one of several things: either the landlords abandoned the idea of trying to construct tenant farming on the estate, the farmers abandoned their rented lands and relocated on better terms on another estate, purchased land for themselves, or people began squatting on one of the many abandoned plantations to be found in the region.

The redefinitions of material space that occurred during the period between 1834 and 1865 expressed the newly emerging spatialities of free labor. Planters attempted to use the new spatial negotiations as a way to control labor while the working population used social space to express their real independence from the plantation economy. These contradictions occasionally resulted in violence, such as the murder of David McLean, and eventually would erupt into an organized rebellion at Morant Bay. This latter event, which serves as the terminal point of the study period, will be touched on in the next, concluding chapter.

Epilogue | 8

The period, however distant, will doubtless arrive when the dominion of the Europeans throughout the American Archipelago, shall no longer exist.
—John Stewart, Jamaican planter, 1808

MORANT BAY

On Saturday, October 7, 1865, a clash between police and a crowd of protestors broke out in the town of Morant Bay, the parochial seat of the parish of St. Thomas-in-the-East. This parish bordered St. David to the east; on a clear day, the coastal town of Morant Bay can be seen from the higher elevations of the eastern Yallahs region. The crowd had assembled outside of the parish courthouse to protest the proceedings against Lewis Miller. Miller was the unfortunate scapegoat caught in the middle of a conflict between the small tenants living and farming on Middleton estate and their landlord, James Williams. The tenants had threatened Williams with a rent strike unless he lowered the rents that he was assessing. Some tenants had already withheld their rent, claiming that the estate was vacant, and thus free for settlement. The tenants generally felt that they were being oppressed by Williams, who was not the landowner of the estate, but a lessee, who was subletting small farms for profit. For his part, Williams threatened his tenants with the possibility of prosecuting them for trespassing. Lewis Miller, whose horse had strayed onto Williams's leased land, was a test case (Heuman, 1994).

Before Miller's case could be tried, an unrelated case was interrupted by James Geoghegan, who loudly protested a fine that was imposed on a youth found guilty of assault. The court ordered Geoghegan's arrest, but the crowd refused to let the police take him into custody. In the ensuing scuffle, at least two policemen were injured. By the time Miller's case was heard, both the police and the crowd were highly agitated. Miller was found guilty and fined 20 shillings; his cousin, a charismatic minister by the name of Paul Bogle, urged Miller to appeal the fine. The crowd dispersed satisfied that, for the time being, Miller had not been arrested (Heuman, 1994).

However, on the following Monday, the court ordered the arrest of 28 people involved in the skirmish with police. On Tuesday, when the police attempted to arrest Bogle in the rural village of Stony Gut, they were immediately surrounded by upwards of 300 armed men. Although some of the officers fled, others were detained and were forced to take an oath of loyalty before being freed. The next day, October 11, Bogle led an organized group of approximately 400 people into Morant Bay, where they engaged in a violent confrontation with the volunteer militia of St. Thomas-in-the-East. The militia, heavily outnumbered, fired into the crowd when a few people began to hurl stones and bottles. In response, the demonstrators rushed the militia, which was forced to retreat into the courthouse. Within minutes, the courthouse and abutting schoolhouse were set on fire, forcing the besieged militia, as well as the parish vestry officials, back into the melee. By the time the peasant forces evacuated Morant Bay, 7 of their number were dead, as were 18 militia and vestrymen (Heuman, 1994).

Heuman has argued that the uprising in Morant Bay was well organized and closely orchestrated by Paul Bogle and his lieutenants. This was not a case of random mob violence, but a planned uprising in resistance to what was perceived as the focus of oppression in this eastern parish: the members of the parish vestry. The town was not looted nor were there random acts of destruction. The only buildings destroyed were the courthouse and schoolhouse, the latter only because it was located next to the court. Although Heuman has suggested that some in the crowd were crying for the deaths of any whites, the evidence suggests that most of those killed in the uprising were targeted. According to Heuman, Bogle's force attacked the vestry because as a body it was the symbol of postemancipation oppression, particularly because its leading members were "involved in disputes with the people over land, justice and wages" (Heuman, 1994:14). As may be expected under a colonial regime, the Morant Bay uprising was answered with a series of brutal reprisals by the authorities. By the end of the incident, over 500 people, primarily black peasants, were dead; George William Gordon, a "colored" member of the House of Assembly, had been court-martialed and executed on spurious charges of planning the uprising; the 200-year-old Jamaica House of Assembly was abolished in favor of Crown Colony government; hundreds of people were publicly flogged; as many as 1000 houses were burned by the British military; the governor of the island (Edward John Eyre) was recalled from Jamaica and eventually tried and acquitted for crimes against the Crown.

In many ways, the uprising at Morant Bay, nestled in the foothills of the Blue Mountains, resulted from the sociospatial dialectic that was in the process of negotiation following emancipation. The planter class had tried to keep the Jamaican peasantry alienated from the land. The freed slaves and their descendants equated land with freedom. Without control over the spatialities of production in, particularly, provision grounds, the people were forced into rent relationships with landlords who extracted rent for profit. Many resisted these efforts, opting out of the system by purchasing or squatting on land. Such strategies were in turn resisted by the elites, who wanted to limit access to space in order to maintain a dependent wage labor force, ready to provide continuous labor when the planters needed it, but left to fend for themselves in off-seasons. Without sounding too apocalyptic or deterministic, it may have been inevitable that clashes between the white landholding class and the primarily black landless and small holding classes would erupt. This dialectic has remained a social force within Jamaica which has been violently expressed in large-scale uprisings like the Morant Bay Rebellion in 1865 and the series of strikes and violent uprisings that occurred in 1938, as well as quotidian, smaller-scale spatial resistance, including squatting and tenant strikes that are still frequent on the island. Resistance to these same spatial dynamics can be read in the work of Marcus Garvey, the philosophy of Rastafarianism, and the political music of Bob Marley and his successors, particularly the dub poets like Mutabaruka. All of these are currently part of the national discourse in Jamaica; all contain elements calling for self-sufficiency for the Jamaican people as individuals and as a nation.

SUMMARY AND CONCLUSION

In summary, the argument presented in this book contends that during a particular period of crisis within the capitalist world system, beginning in the last decades of the eighteenth century, and eventually resulting in the abolition of slavery, members of the regional elite classes in Jamaica attempted to maintain and reinforce their social position, in part, by redefining material, cognitive, and social spaces on the island. Specifically, this study examined the spatial manipulations surrounding coffee production in the Yallahs drainage in the Blue Mountains.

I have argued that periods of crisis are endemic within the political economy of capitalism. These fluctuations in the logic of the system occasionally result in the restructuring of the logic of accumulation as the relative prosperity of geographic regions or economic sectors wanes or waxes. I have further argued that observers of the eighteenth and nineteenth centuries perceived a crisis gripping the political economy of Jamaica as early as 1774; from that date a series of plantation theorists offered suggestions in the colonial discourse on how to ameliorate the perceived crises by breaking the hegemony of sugar production and thus diversifying the economy of the island; by creating new productive spaces, and thus exerting European control over the sparsely inhabited interior districts of Jamaica; by alleviating demographic manifestations of crises by implementing a more rational labor management system, although worked by slaves, through a more humane allocation of space by slave masters on their plantations; and by spatially defining coffee plantations.

I have also argued that there were two periods of transition resulting from this crisis, during which spatial manipulations were used to create and reinforce systems of oppression, and were in turn resisted. The first was characterized by the introduction of coffee in the decades prior to emancipation; the second by the renegotiation of social and spatial relationships in the decades following the abolition of slavery. Using documentary, cartographic, and archaeological data, I have interpreted the spatial dynamics of these phases of Jamaican history.

In the first of these periods, coffee was introduced as a large-scale export commodity. New spaces, defined as plantations, and based on a cognized model influenced by the existing spatial relationships on Jamaican sugar plantations, and the advice of emigré French planters from nearby St. Domingue, were created in the Yallahs region. During this phase of Jamaican history, colonialists preconceived, and to a degree successfully followed, a cognitive plan of designing space as one method by which to adapt the slave labor system to coffee production. These plantations flourished for several decades, until faced with a new element within the reorganization of capital in the British colonial world: the abolition of slavery

Abolition came in two phases, apprenticeship followed by "full freedom." During these phases of postslavery Jamaican capitalism, the spatial logic of coffee plantations was redefined by the elites, in the hope of creating a peaceful transition to a wage-based system of labor exploitation. However, these manipulations were resisted, as the African Jamaican population had developed spatialities, particularly of their houses and provision grounds, that were significantly different from the cognitive models espoused by the planter class. To the African Jamaican people, to be free was to exert control over their own spatialities, and to control their own spaces of production. The schemes proposed by the planters never fully materialized, as dialectical conflict erupted as the planters attempted to alienate labor from both the means of production and land, and the laborers attempted to exert increasing amounts of control over their own spaces, including purchasing plots of land on which they hoped to become self-sufficient producers.

While I do not mean to be so bold as to propose anything remotely like a new paradigm or research program for historical archaeology (Orser, 1996), I do hope that this study has demonstrated how landscape archaeology can contribute to our understanding of the global political economy that emerged during the nineteenth century. Archaeology can be many things to many people; here I have hoped to demonstrate how historical archaeology can be understood as being the study of the material culture of capitalism. Specifically, I have presented what I think is a practical way to understand space archaeologically. In considering spatial phenomena to be simultaneously experienced materially, cognitively, and socially, I believe that we can better connect the material record we excavate as archaeologists to the more abstract phenomena we experience as people. In doing so, I have hoped to provide a practical way that data can be linked to theory, and thus how historical archaeology can contribute to our understanding of the important role material culture plays in the negotiation of the political economy under which we still operate.

References

Adams, W. H., 1976, Trade Networks and Interaction Spheres: A View from Silcott. *Historical Archaeology* 10:99–111.

Adams, W. H., 1977, *Silcott, Washington: Ethnoarchaeology of a Rural American Community*. Laboratory of Anthropology, Washington State University, Pullman.

Adams, W. H., and Boling, S. J., 1989, Status and Ceramics for Planters and Slaves on Three Georgia Coastal Plantations. *Historical Archaeology* 23(1):69–96.

Anstey, R., 1968, Capitalism and Slavery: A Critique. *Economic History Review* 21(2): 307–320.

Anstey, R., 1975, *The Atlantic Slave Trade and British Abolition, 1760–1810*. Macmillan, London.

Anstey, R., 1980, The Pattern of British Abolitionism in the Eighteenth and Nineteenth Centuries. In *Anti-Slavery, Religion and Reform*, edited by C. Bolt and S. Drescher, pp. 19–42. William Davidson and Sons, Kent.

Armstrong, D. V., 1982, The "Old Village" at Drax Hall: An Archaeological Progress Report. *Journal of New World Archaeology* 5(2):87–103.

Armstrong, D. V., 1985, An Afro-Jamaican Slave Settlement: Archaeological Investigations at Drax Hall. In *The Archaeology of Slavery and Plantation Life*, edited by T. Singleton, pp. 261–287. Academic Press, New York.

Armstrong, D. V., 1990, *The Old Village and the Great House: An Archaeological and Historical Examination of Drax Hall Plantation, St. Ann's Bay, Jamaica*. University of Illinois Press, Urbana.

Armstrong, D. V., and Kelly, K. G., 1990, Settlement Pattern Shifts in a Jamaican Slave Village, Seville Estate, St. Ann's Bay, Jamaica. Paper presented at the 23rd Annual Conference of the Society for Historical Archaeology, Tucson, AZ.

Armstrong, D. V., and Kelly, K. G., 1991, Processes of Change and Patterns of Meaning in a Jamaican Slave Village. Paper presented at the 24th Annual Conference of the Society for Historical Archaeology, Richmond, VA.

Babson, D. W., 1990, The Archaeology of Racism and Ethnicity on Southern Plantations. *Historical Archaeology* 24(4):20–28.

Bakan, A. B., 1990, *Ideology and Class Conflict in Jamaica: The Politics of Rebellion*. McGill-Queen's University Press, Montreal.

Baram, U., 1989, *"Boys, Be Ambitious": Landscape Manipulation in Nineteenth-Century Western Massachusetts*. Master's thesis, University of Massachusetts, Amherst.

Barclay, A., 1969 [1828], *A Practical View of the Present State of Slavery in the West Indies*. Mnemosyne Publishing Company, Miami.

Beaudry, M. C., 1989, The Lowell Boott Mills Complex and Its Housing: Material Expressions of Corporate Ideology. *Historical Archaeology* 23(1):19–32.

Beaudry, M. C., 1996, Why Gardens? In *Landscape Archaeology: Reading and Interpreting the American Historical Landscape*, edited by R. Yamin and K. B. Metheny, pp. 3–5. University of Tennessee Press, Knoxville.

Beaudry, M. C., Cook, L., and Mrozowski, S. A., 1991, Artifacts and Active Voices: Material Culture as Social Discourse. In *The Archaeology of Inequality*, edited by R. H. McGuire and R. Paynter, pp. 150–191. Basil Blackwell, Oxford.

Beaudry, M. C., and Mrozowski, S. A., 1987a, *Interdisciplinary Investigations of the Boott Mills, Lowell, Massachusetts, Vol. I: Life at the Boarding Houses*. Division of Cultural Resources, North Atlantic Region, National Park Service, Boston.

Beaudry, M. C., and Mrozowski, S. A., 1987b, *Interdisciplinary Investigations of the Boott Mills, Lowell, Massachusetts, Vol. II: The Kirk Street Agent's House*. Division of Cultural Resources, North Atlantic Region, National Park Service, Boston.

Beckford, W., 1790, *A Descriptive Account of the Island of Jamaica*. T. and J. Egerton, London.

Berlin, I., and Morgan, P. D., 1993, Labor and the Shaping of Slave Life in the Americas. In *Cultivation and Culture: Labor and the Shaping of Slave Life in the Americas*, edited by I. Berlin and P. D. Morgan, pp. 1–45. University Press of Virginia, Charlottesville.

Bigelow, J., 1970 [1851], *Jamaica in 1850; or, the Effects of Sixteen Years of Freedom on a Slave Colony*. Negro Universities Press, Westport.

Blackburn, R., 1988, *The Overthrow of Colonial Slavery, 1776–1848*. Verso, New York.

Bodley, J., 1990, *Victims of Progress*, 3rd edition. Mayfield, Mountain View.

Bonner, T., 1974, Blue Mountain Expedition: Exploratory Excavations at Nanny Town by the Scientific Exploration Society. *Jamaica Journal* 8(2–3):46–50.

Braudel, F., 1984, *The Perspective of the World, Vol. 3: Civilization and Capitalism 15th–18th Century*. Harper & Row, New York.

Butler, K. M., 1995, *The Economics of Emancipation: Jamaica & Barbados, 1823–1843*. University of North Carolina Press, Chapel Hill.

Carson, C., Barka, N. F., Kelso, W., Wheeler, G., and Upton, D., 1981, Impermanent Architecture in the Southern American Colonies. *Winterthur Portfolio* 16(2/3):135–196.

Checkland, S. G., 1957, Finance for the West Indies, 1780–1815. *Economic History Review* 10:461–469.

Collins, D., 1971 [1811], *Practical Rules for the Management and Medical Treatment of Negro Slaves, in the Sugar Colonies, by a Professional Planter*. Books for Libraries Press, Freeport.

Costello, J., 1991, *Variability and Economic Change in the California Missions: An Historical and Archaeological Study*. Ph.D. dissertation, University of California, Santa Barbara.

Cotter, C. S., 1970, Sevilla Nueva: The Story of an Excavation. *Jamaica Journal* 4(2): 15–22.

Craton, M., 1978, *Searching for the Invisible Man: Slaves and Plantation Life in Jamaica*. Harvard University Press, Cambridge, MA.

Craton, M., and Walvin, J., 1970, *A Jamaican Plantation: The History of Worthy Park, 1670–1970*. W. H. Allen, New York.

Davis, D. B., 1975, *The Problem of Slavery in the Age of Revolution, 1770–1823*. Cornell University Press, Ithaca.

Deagan, K., 1983, *Spanish St. Augustine: The Archaeology of a Colonial Creole Community*. Academic Press, San Diego.

De Corse, C., 1991, West African Archaeology and the Atlantic Slave Trade. *Slavery and Abolition* 12(2):92–96.

Deerr, N., 1950, *The History of Sugar*. Chapman and Hall, London.

Deetz, J., 1977, *In Small Things Forgotten: An Archaeology of Early American Life*. Doubleday, New York.

Deetz, J., 1991, Archaeological Evidence of Sixteenth- and Seventeenth-Century Encounters. In *Historical Archaeology in Global Perspective*, edited by L. Falk, pp. 1–10. Smithsonian Institution Press, Washington, DC.

Deetz, J., 1993, *Flowerdew Hundred*. University of Virginia Press, Charlottesville.

Delle, J. A., 1989, *A Spatial Analysis of Sugar Plantations on St. Eustatius, Netherlands Antilles*. Master's thesis, College of William and Mary, Williamsburg.

Delle, J. A., 1994, A Spatial Analysis of Sugar Plantations on St. Eustatius, Netherlands Antilles. In *Spatial Patterning in Historical Archaeology: Selected Studies of Settlement*, edited by D. W. Linebaugh and G. G. Robinson, pp. 33–62. King and Queen Press, Williamsburg.

Delle, J. A., 1995, Space, Power and the Transition to Free Labor in the Coffee Plantation Economy of Jamaica, 1800–1850. Paper presented at the 28th Annual Conference of the Society for Historical Archaeology, Washington, DC.

Delle, J. A., 1996, *An Archaeology of Crisis: The Manipulation of Social Spaces in the Blue Mountains of Jamaica, 1790–1865*. Ph.D. dissertation, University of Massachusetts, Amherst.

Delle, J. A., in press, Extending Europe's Grasp: An Archaeological Comparison of Colonial Spatial Processes in Ireland and Jamaica. In *Archaeology of the British, 1600–1800: Views from Two Worlds. Proceedings of the Special Joint Conference of the Society for Historical Archaeology and the Society for Post-medieval Archaeology*. David Brown Publishers, London.

Delle, J. A., Leone, M. P., and Mullins, P., in press, The Historical Archaeology of the Modern State. In *The Routledge Companion Encyclopedia of Archaeology*, edited by G. Barker and A. Grant. Routledge, London.

Dethlefsen, E., 1982, The Historical Archaeology of St. Eustatius. *Journal of New World Archaeology* 5(2):73–86.

Drescher, S., 1977, *Econocide: British Slavery in the Era of Abolition*. University of Pittsburgh Press, Pittsburgh.

Dunn, R., 1972, *Sugar and Slaves*. University of North Carolina Press, Chapel Hill.

Edwards, B., 1810, *The History Civil and Commercial of the British Colonies in the West Indies*. Levis and Weaver, Philadelphia.

Engerman, S., and Eltis, D., 1980, Economic Aspects of the Abolition Debate. In *Anti-Slavery, Religion and Reform*, edited by C. Bolt and S. Drescher, pp. 149–162. William Davidson and Sons, Kent.

Epperson, T. W., 1990, Race and the Disciplines of the Plantation. *Historical Archaeology* 24(4):29–36.

Falk, L. (Ed.), 1991, *Historical Archaeology in Global Perspective*. Smithsonian Institution Press, Washington, DC.

Farnsworth, P., 1987, *The Economics of Acculturation in the California Missions: A Historical and Archaeological Study of Mission Nuestra Senora de la Soledad*. Ph.D. dissertation, University of California, Los Angeles.

Fee, J. M., 1993, Idaho's Chinese Gardens. In *Hidden Heritage: Historical Archaeology of the Overseas Chinese*, edited by P. Wegars, pp. 65–96. Baywood Publishing Company, Amityville, NY.

Ferguson, L. E., 1980, Looking for the "Afro" in Colono-Indian Pottery. In *Archaeological Perspectives on Ethnicity in America: Afro-American and Asian American Culture History*, edited by R. Schuyler, pp. 14–28. Baywood, Farmingdale, NY.

Ferguson, L. E., 1992, *Uncommon Ground: Archaeology and Colonial African-America*. Smithsonian Institution Press, Washington, DC.

Fitts, R. K., 1996, The Landscapes of Northern Bondage. *Historical Archaeology* 30(2): 54–73.

Foucault, M., 1979, *Discipline and Punish: The Birth of the Prison*. Vintage Books, New York.

Garman, J. C., 1994, Viewing the Color Line through the Material Culture of Death. *Historical Archaeology* 28(3):74–93.

Geggus, D., 1993, Sugar and Coffee Cultivation in Saint Domingue and the Shaping of the Slave Labor Force. In *Cultivation and Culture: Labor and the Shaping of Slave Life in the Americas*, edited by I. Berlin and P. D. Morgan, pp. 73–98. University of Virginia Press, Charlottesville.

Gibb, J. A., and King, J. A., 1991, Gender, Activity Areas, and Homelots in the 17th-Century Chesapeake Region. *Historical Archaeology* 25(4):109–131.

Glassie, H., 1975, *Folk Housing in Middle Virginia: A Structural Analysis of Historic Artifacts*. University of Tennessee Press, Knoxville.

Gordon, D. M., Edwards, R., and Reich, M., 1982, *Segmented Work, Divided Workers: The Historical Transformations of Labor in the U.S.* Cambridge University Press, London.

Green, W., 1976, *British Slave Emancipation: The Sugar Colonies and the Great Experiment, 1830–1865*. Clarendon Press, Oxford.

Habermas, J., 1975, *Legitimation Crisis*. Beacon Press, Boston.

Hall, D., 1959, *Free Jamaica, 1838–1865: An Economic History*. Yale University Press, New Haven.

Hall, D., 1989, *In Miserable Slavery: Thomas Thistlewood in Jamaica, 1750–86*. Macmillan, London.

Hall, M., 1992, Small Things and the Mobile, Conflictual Fusion of Power, Fear and Desire. In *The Art and Mystery of Historical Archaeology*, edited by A. Yentsch and M. Beaudry, pp. 373–399. CRC Press, Boca Raton.

Handler, J., 1996, A Prone Burial from a Plantation Slave Cemetery in Barbados, West Indies: Possible Evidence for an African-Type Witch or Other Negatively Viewed Person. *Historical Archaeology* 30(3):76–86.

Handler, J., and Lange, F. W., 1978, *Plantation Slavery In Barbados: An Archaeological and Historical Investigation*. Harvard University Press, Cambridge, MA.

Hardesty, D. L., 1988, *The Archaeology of Mining and Miners: A View from the Silver State*. Society for Historical Archaeology, California, PA.

Hardesty, D. L., 1994, Class, Gender Strategies, and Material Culture in the Mining West. In *Those of Little Note: Gender, Race, and Class in Historical Archaeology*, edited by E. Scott, pp. 129–145. University of Arizona Press, Tucson.

Harley, J. B., 1988, Silences and Secrecy: The Hidden Agenda of Cartography in Early Modern Europe. *Imago Mundi* 40:57–76.

Harley, J. B., 1994, New England Cartography and the Native Americans. In *American Beginnings: Exploration, Culture, and Cartography in the Land of Norumbega*, edited by E. Barker, E. A. Churchill, R. d'Abate, K. L. Jones, V. A. Konrad, and H. E. L. Prins, pp. 287–313. University of Nebraska Press, Lincoln.

Harrington, F., 1989, The Emergent Elite in Early 18th Century Portsmouth Society: The Archaeology of the Joseph Sherburne Houselot. *Historical Archaeology* 23(1):2–18.

Harrison, B., and Bluestone, B., 1988, *The Great U-Turn: Corporate Restructuring and the Polarizing of America*. Basic Books, New York.

Harvey, D., 1982, *The Limits to Capital*. University of Chicago Press, Chicago.

Harvey, D., 1990, *The Condition of Post-Modernity: An Enquiry into the Origins of Cultural Change*. Basil Blackwell, Oxford.

Hautaniemi, S., 1992, The W.E.B. Du Bois Site: Material Culture and the Creation of Race and Gender. Paper presented at the 25th Annual Conference of the Society for Historical Archaeology, Kingston, Jamaica.

Heuman, G., 1981, *Between Black and White: Race, Politics and the Free Coloreds in Jamaica, 1792–1865*. Greenwood Press, Westport.

Heuman, G., 1994, *"The Killing Time": The Morant Bay Rebellion in Jamaica*. University of Tennessee Press, Knoxville.

Higman, B., 1976, *Slave Population and Economy in Jamaica, 1807–1834*. Cambridge University Press, London.

Higman, B., 1984, *Slave Populations of the British Caribbean*. Johns Hopkins University Press, Baltimore.

Higman, B., 1986, Jamaican Coffee Plantations 1780–1860: A Cartographic Analysis. *Caribbean Geography* 2:73–91.

Higman, B., 1988, *Jamaica Surveyed: Plantation Maps and Plans of the Eighteenth and Nineteenth Centuries*. Institute of Jamaica, Kingston.

Hodder, I., 1985, Postprocessual Archaeology. In *Advances in Archaeological Method and Theory*, Vol. 8, edited by M. B. Schiffer, pp. 1–26. Academic Press, New York.

Holt, T., 1992, *The Problem of Freedom: Race, Labor, and Politics in Jamaica and Britain, 1832–1938*. Johns Hopkins University Press, Baltimore.

Hood, J. E., 1995, Some Observations on Interpreting the Archaeology of a New England Village to the Public. Paper presented at the 60th annual meeting of the Society for American Archaeology, Minneapolis.

Hood, J. E., 1996, Social Relations and the Cultural Landscape. In *Landscape Archaeology: Reading and Interpreting the American Historical Landscape*, edited by R. Yamin and K. B. Metheny, pp. 121–146. University of Tennessee Press, Knoxville.

Hood, J. E., and Reinke, R., 1989, Manipulating the Landscape: Some Evidence from Deerfield, Massachusetts. Paper read at the annual meeting of the Eastern States Archaeological Federation, East Windsor, CT.

Hood, J. E., and Reinke, R., 1990, *Report on the Hadley Palisade Project*. Hadley Historical Commission, Hadley.

Howson, J., 1990, Social Relations and Material Culture: A Critique of the Archaeology of Plantation Slavery. *Historical Archaeology* 24(4):78–91.

Howson, J., 1995, *Colonial Goods and the Plantation Village: Consumption and the Internal Economy in Montserrat from Slavery to Freedom*. Ph.D. dissertation, New York University, New York.

Hudgins, C. L., 1990, Robert "King" Carter and the Landscape of Tidewater Virginia in the Eighteenth Century. In *Earth Patterns: Essays in Landscape Archaeology*, edited by W. Kelso and R. Most, pp. 59–70. University Press of Virginia, Charlottesville.

James, C. L. R., 1963, *The Black Jacobins: Toussaint L'Ouverture and the San Domingo Revolution*. Vintage, New York.

Jamieson, R., 1996, *The Domestic Architecture and Material Culture of Colonial Cuenca, Ecuador, AD 1600–1800*. Ph.D. dissertation, Department of Archaeology, University of Calgary, Calgary.

Jennings, F., 1988, *Empire of Fortune: Crown, Colonies, and Tribes in the Seven Years War in America*. Norton, New York.

Jennings, L., 1988, *French Reaction to British Slave Emancipation*. LSU Press, Baton Rouge.

Johnson, M., 1991, Enclosure and Capitalism: The History of a Process. In *Processual and Postprocessual Archaeologies: Multiple Ways of Knowing the Past*, edited by R. Preucel, pp. 159–67. Center for Archaeological Investigations, Carbondale.

Johnson, M., 1993, *Housing Culture: Traditional Architecture in an English Landscape*. Smithsonian Institution Press, Washington, DC.

Johnson, M., 1996, *An Archaeology of Capitalism*. Blackwell Publishers, Cambridge, MA.

Joseph, J. W., 1995, Los Cafecultores de Maraguez: A Social and Technological History of Coffee Processing in the Cerrillos Valley, Municipio Ponce, Puerto Rico. Paper read at the 60th annual meeting of the Society for American Archaeology, Minneapolis.

Joseph, J. W., Ramos y Ramirez de Arellano, A., and Pabon de Rocafort, A., 1987, *Los Caficultores de Maraguez: An Architectural and Social History of Coffee Processing in the Cerrillos Valley, Ponce, Puerto Rico*. Garrow and Associates, Atlanta.

Kelly, K. G., 1995, *Transformation and Continuity in Savi, a West African Trade Town: An Archaeological Investigation of Culture Change on the Coast of Benin During the 17th and 18th Centuries*. Ph.D. dissertation, University of California, Los Angeles.

Kelly, K. G., 1997, The Archaeology of African–European Interaction: Investigating the Social Roles of Trade, Traders, and the Use of Space in the Seventeenth- and Eighteenth-Century Hueda Kingdom, Republic of Benin. *World Archaeology* 28(3):351–369.

Kelso, W. M., 1984, *Kingsmill Plantation, 1619–1800: Archaeology of Country Life in Colonial Virginia*. Academic Press, Orlando.

Kelso, W. M., 1989, Comments on the 1987 Society for Historical Archaeology Landscape Symposium. *Historical Archaeology* 23(2):48–49.

Kelso, W. M., 1990, Landscape Archaeology at Thomas Jefferson's Monticello. In *Earth Patterns: Essays in Landscape Archaeology*, edited by W. Kelso and R. Most, pp. 7–22. University Press of Virginia, Charlottesville.

Kelso, W. M., and Most, R. (Eds.), 1990, *Earth Patterns: Essays in Landscape Archaeology*. University of Virginia Press, Charlottesville.

Kondratieff, N. D., 1979, The Long Waves in Economic Life. *Review* 2:519–562.

Kryder-Reid, E., 1994, "As Is the Gardener, So Is the Garden": The Archaeology of Landscape as Myth. In *The Historical Archaeology of the Chesapeake*, edited by P. A. Shackel and B. J. Little, pp. 131–148. Smithsonian Institution Press, Washington, DC.

Laborie, P. J., 1798, *The Coffee Planter of Saint Domingo*. T. Caddell and W. Davies, London.

LeeDecker, C. H., Klein, T. H., Holt, C. A., and Friedlander, A., 1987, Nineteenth-Century Households and Consumer Behavior in Wilmington, Delaware. In *Consumer Choice in Historical Archaeology*, edited by S. M. Spencer-Wood, pp. 233–259. Plenum Press, New York.

Lefebvre, G., 1947, *The Coming of the French Revolution*. Princeton University Press, Princeton.

Lefebvre, H., 1991, *The Production of Space*. Basil Blackwell, Oxford.

Leone, M. P., 1984, Interpreting Ideology in Historical Archaeology: The William Paca Garden in Annapolis, Maryland. In *Ideology, Power and Prehistory*, edited by D. Miller and C. Tilley, pp. 25–35. Cambridge University Press, London.

Leone, M. P., 1988a, The Georgian Order as the Order of Merchant Capitalism in Annapolis, Maryland. In *The Recovery of Meaning: Historical Archaeology in the Eastern United States*, edited by M. P. Leone and P. B. Potter, Jr., pp. 219–229. Smithsonian Institution Press, Washington, DC.

Leone, M. P., 1988b, The Relationship Between Archaeological Data and the Documentary Record: 18th Century Gardens in Annapolis, Maryland. *Historical Archaeology* 22(1):29–35.

Leone, M. P., 1989, Issues in Historic Landscapes and Gardens. *Historical Archaeology* 23(1):45–47.

Leone, M. P., 1995, A Historical Archaeology of Capitalism. *American Anthropologist* 97(2):251–268.

Leone, M. P., and Potter, P. B., Jr., 1988, Introduction: Issues in Historical Archaeology. In *The Recovery of Meaning: Historical Archaeology in the Eastern United States*, edited by M. P. Leone and P. B. Potter, Jr., pp. 1–20. Smithsonian Institution Press, Washington, DC.

Leone, M. P., and Shackel, P. A., 1990, Plane and Solid Geometry in Colonial Gardens in Annapolis, Maryland. In *Earth Patterns: Essays in Landscape Archaeology*, edited by W. Kelso and R. Most, pp. 153–168. University Press of Virginia, Charlottesville.

Leone, M. P., and Silberman, N. A. (Eds.), 1995, *Invisible America: Unearthing Our Hidden History*. Henry Holt, New York.

Lewis, K., 1984, *The American Frontier: An Archaeological Study of Settlement Pattern and Process*. Academic Press, New York.

Lewis, K., 1985, Plantation Layout and Function in the South Carolina Lowcountry. In *The Archaeology of Slavery and Plantation Life*, edited by T. Singleton, pp. 35–66. Academic Press, San Diego.

Lewis, M., 1929 [1834], *Journal of a West Indian Proprietor, 1815–17. Edited with an Introduction by Mona Wilson*. Houghton Mifflin, Boston.

Little, B., 1994, "She Was...an Example to Her Sex": Possibilities for a Feminist Historical Archaeology. In *Historical Archaeology of the Chesapeake*, edited by P. A. Shackel and B. J. Little, pp. 189–204. Smithsonian Institution Press, Washington, DC.

Lobdell, R., 1972, Patterns of Investment and Sources of Credit in the British West Indian Sugar Industry. *Journal of Caribbean History* 4:31–53.

Long, E., 1970 [1774], *The History of Jamaica*. Frank Cass & Co., London.

Lucketti, N., 1990, Archaeological Excavations at Bacon's Castle, Surry County, Virginia. In *Earth Patterns: Essays in Landscape Archaeology*, edited by W. Kelso and R. Most, pp. 23–42. University Press of Virginia, Charlottesville.

Mangan, P., 1994, *Changes in the Landscape During the Transition from Feudalism to Capitalism: A Case Study of Montblanc, Catalonia, Spain*. Ph.D. dissertation, University of Massachusetts, Amherst.

Marx, K., 1967, *Capital: A Critique of Political Economy*, Vol. 1. International Publishers, New York.

McGuire, R. H., 1988, Dialogues with the Dead: Ideology and the Cemetery. In *The Recovery of Meaning: Historical Archaeology in the Eastern United States*, edited by M. P. Leone and P. B. Potter, Jr., pp. 435–480. Smithsonian Institution Press, Washington, DC.

McGuire, R. H., 1991, Building Power in the Cultural Landscape of Broome County, New York, 1880 to 1940. In *The Archaeology of Inequality*, edited by R. H. McGuire and R. Paynter, pp. 102–124. Basil Blackwell, Oxford.

McGuire, R. H., 1992, *A Marxist Archaeology*. Academic Press, San Diego.

McGuire, R. H., 1993, Archaeology and Marxism. In *Archaeological Method and Theory*, Vol. 5, edited by M. Schiffer, pp. 101–157. University of Arizona Press, Tucson.

McKee, L., 1996, The Archaeology of Rachel's Garden. In *Landscape Archaeology: Reading and Interpreting the American Historical Landscape*, edited by R. Yamin and K. B. Metheny, pp. 70–90. University of Tennessee Press, Knoxville.

Metheny, K. B., Kratzer, J., Yentsch, A. E., and Goodwin, C. M., 1996, Method in Landscape Archaeology: Research Strategies in a Historic New Jersey Garden. In *Landscape Archaeology: Reading and Interpreting the American Historical Landscape*, edited by R. Yamin and K. B. Metheny, pp. 6–31. University of Tennessee Press, Knoxville.

Miller, D., 1987, *Material Culture and Mass Consumption*. Basil Blackwell, Oxford.

Miller, D., and Tilley, C., 1984, Ideology, Power and Prehistory: An Introduction. In *Ideology, Power and Prehistory*, edited by D. Miller and C. Tilley, pp. 1–15. Cambridge University Press, London.

Miller, N. F., and Gleason, K. L. (Eds.), 1994, *The Archaeology of Garden and Field*. University of Pennsylvania Press, Philadelphia.

Mintz, S., 1961, The Question of Caribbean Peasantries: A Comment. *Caribbean Studies* 1:31–34.

Mintz, S., 1979, Time, Sugar, and Sweetness. *Marxist Perspectives* 8:56–73.

Mintz, S., 1985a, *Sweetness and Power: The Place of Sugar in Modern History*. Viking, New York.

Mintz, S., 1985b, From Plantations to Peasantries in the Caribbean. In *Caribbean Contours*, edited by S. Mintz and S. Price, pp. 127–153. Johns Hopkins University Press, Baltimore.

Mintz, S., and Hall, D., 1960, *The Origins of the Jamaican Internal Marketing System*. Yale University Publications in Anthropology, New Haven.

Montieth, K., 1991, *The Coffee Industry in Jamaica, 1790–1850*. Master's thesis, University of the West Indies at Mona, Kingston.

Moore, S. M., 1985, Social and Economic Status on the Coastal Plantation: An Archaeological Perspective. In *The Archaeology of Slave and Plantation Life*, edited by T. Singleton, pp. 141–162. Academic Press, San Diego.

Morrissey, M., 1989, *Slave Women in the New World: Gender Stratification in the Caribbean*. University of Kansas Press, Lawrence.

Mrozowski, S. A., 1987, Exploring New England's Evolving Urban Landscape. In *Living in Cities*, edited by E. Staski, pp. 1–9. Society for Historical Archaeology, Ann Arbor.

Mrozowski, S. A., 1991, Landscapes of Inequality. In *The Archaeology of Inequality*, edited by R. H. McGuire and R. Paynter, pp. 79–101. Basil Blackwell, Oxford.

Mrozowski, S. A., Ziesing, G. H., and Beaudry, M. C., 1996, *Living on the Boott: Historical Archaeology at the Boott Mills Boardinghouses, Lowell, Massachusetts*. University of Massachusetts Press, Amherst.

Muller, N. L., 1994, The House of the Black Burghardts: An Investigation of Race, Gender, and Class at the W. E. B. Du Bois Boyhood Homesite. In *Those of Little Note: Gender, Race, and Class in Historical Archaeology*, edited by E. Scott, pp. 81–94. University of Arizona Press, Tucson.

Mullins, P. R., 1996, *The Contradictions of Consumption: An Archaeology of African America and Consumer Culture, 1850–1930*. Ph.D. dissertation, University of Massachusetts, Amherst.

Neiman, F., 1978, Domestic Architecture at the Clifts Plantation: The Social Context of Early Virginia Building. *Northern Neck of Virginia Historical Magazine*, 3096–3128.

Noël Hume, A., 1974, *Archaeology and the Colonial Gardener*. Colonial Williamsburg Foundation, Williamsburg.

Noël Hume, I., 1982, *Martin's Hundred*. Knopf, New York.

O'Brien, M., and Majewski, T., 1989, Wealth and Status in the Upper South Socioeconomic System of Northeastern Missouri. *Historical Archaeology* 23(2):60–95.

O'Connor, J., 1984, *Accumulation Crisis*. Basil Blackwell, Oxford.

O'Connor, J., 1987, *The Meaning of Crisis*. Basil Blackwell, Oxford.

Orser, C. E., Jr., 1988a, The Archaeological Analysis of Plantation Society: Replacing Status and Caste with Economics and Power. *American Antiquity* 53(4):735–751.

Orser, C. E., Jr., 1988b, *The Material Basis of the Postbellum Tenant Plantation: Historical Archaeology in the South Carolina Piedmont*. University of Georgia Press, Athens.

Orser, C. E., Jr., 1988c, Toward a Theory of Power for Historical Archaeology: Plantations and Space. In *The Recovery of Meaning: Historical Archaeology in the Eastern United States*, edited by M. P. Leone and P. B. Potter, Jr., pp. 235–262. Smithsonian Institution Press, Washington, DC.

Orser, C. E., Jr., 1990, Archaeological Approaches to New World Plantation Slavery. In *Archaeological Method and Theory*, edited by M. B. Schiffer, pp. 111–154. University of Arizona, Tucson.

Orser, C. E., Jr., 1991, The Continued Pattern of Dominance: Landlord and Tenant on the Postbellum Cotton Plantation. In *The Archaeology of Inequality*, edited by R. H. McGuire and R. Paynter, pp. 40–54. Basil Blackwell, Oxford.

Orser, C. E., Jr., 1992, *In Search of Zumbi: Preliminary Archaeological Research at the Serra da Berriga, State of Alagoas, Brazil*. Midwestern Archaeological Research Center, Illinois State University, Normal.

Orser, C. E., Jr., 1996, *A Historical Archaeology of the Modern World*. Plenum Press, New York.

Orser, C. E., Jr., and Nekola, A. M., 1985, Plantation Settlement from Slavery to Tenancy: An Example from a Piedmont Plantation in South Carolina. In *The Archaeology of Slavery and Plantation Life*, edited by T. Singleton, pp. 67–94. Academic Press, San Diego.

Otto, J. S., 1977, Artifacts and Status Differences: A Comparison of Ceramics for Planter, Overseer, and Slave Sites on an Antebellum Plantation. In *Research Strategies in Historical Archaeology*, edited by S. South, pp. 91–118. Academic Press, New York.

Otto, J. S., 1984, *Canon's Point Plantation 1794–1860: Living Conditions and Status Patterns in the Old South*. Academic Press, New York.

Pares, R., 1950, *A West-India Fortune*. Longmans, Green and Co., New York.

Pastron, A. G., and Hattori, E. M. (Eds.), 1990, *The Hoff Store Site and Gold Rush Merchandise From San Francisco, California*. Society for Historical Archaeology, Pleasant Hill, CA.

Paterson, R., 1843, *Remarks on the Present State of Cultivation in Jamaica; the Habits of the Peasantry; and Remedies Suggested for the Improvement of Both*. W. Burness, Edinburgh.

Patterson, T., 1993, *Archaeology: The Historical Development of Civilizations*, 2nd edition. Prentice–Hall, Englewood Cliffs.

Paynter, R., 1980, *Long Distance Processes, Stratification, and Settlement Pattern: An Archaeological Perspective*. Ph.D. dissertation, Department of Anthropology, University of Massachusetts, Amherst.

Paynter, R., 1981, Social Complexity in Peripheries: Problems and Models. In *Archaeological Approaches to the Study of Complexity*, edited by S. E. van der Leeuw, pp. 118–141. A. E. van Giffen Institute, Amsterdam.

Paynter, R., 1982, *Models of Spatial Inequality*. Academic Press, New York.

Paynter, R., 1983, Expanding the Scope of Settlement Analysis. In *Archaeological Hammers and Theories*, edited by J. Moore and A. Keene, pp. 234–277. Academic Press, New York.

Paynter, R., 1985, Surplus Flow between Frontiers and Homelands. In *The Archaeology of Frontiers and Boundaries*, edited by S. W. Green and S. M. Perlman, pp. 163–211. Academic Press, Orlando.

Paynter, R., 1988, Steps to an Archaeology of Capitalism: Material Change and Class Analysis. In *The Recovery of Meaning: Historical Archaeology in the Eastern United States*, edited by M. P. Leone and P. B. Potter, Jr., pp. 407–422. Smithsonian Institution Press, Washington, DC.

Paynter, R., 1989, The Archaeology of Equality and Inequality. *Annual Review of Anthropology* 18:369–399.

Paynter, R., 1990, Afro-Americans in the Massachusetts Historical Landscape. In *The Politics of the Past*, edited by P. Gathercole and D. Lowenthal, pp. 49–62. Unwin Hyman, London.

Paynter, R., Hautaniemi, S., and Muller, N. L., 1994, The Landscapes of the W.E.B. Du Bois Boyhood Homesite: An Agenda for an Archaeology of the Color Line. In *Race*, edited by S. Gregory and R. Sanjek, pp. 285–318. Rutgers University Press, New Brunswick.

Paynter, R., and McGuire, R. H., 1991, The Archaeology of Inequality: Material Culture, Domination and Resistance. In *The Archaeology of Inequality*, edited by R. H. McGuire and R. Paynter, pp. 1–27. Basil Blackwell, Oxford.

Perry, W. R., 1996, *Archaeology of the Mfecane / Difaqane: Landscape Transformations in Post-15th-Century Southern Africa*. Ph.D. dissertation, City University of New York, New York.

Pogue, D., 1996, Giant in the Earth: George Washington, Landscape Designer. In *Landscape Archaeology: Reading and Interpreting the American Historical Landscape*, edited by R. Yamin and K. B. Metheny, pp. 52–69. University of Tennessee Press, Knoxville.

Praetzellis, A., and Praetzellis, M., 1989, "Utility and Beauty Should Be One": The Landscape of Jack London's Ranch of Good Intentions. *Historical Archaeology* 23(1):33–44.

Pulsipher, L., 1994, The Landscapes and Ideational Roles of Caribbean Slave Gardens. In *The Archaeology of Garden and Field*, edited by N. Miller and K. Gleason, pp. 202–221. University of Pennsylvania Press, Philadelphia.

Pulsipher, L., and Goodwin, C., 1982, *Galways: A Caribbean Sugar Plantation — A Report of the 1981 Field Season.* Department of Archaeology, Boston University, Boston.

Purser, M., 1991, "Several Paradise Ladies Are Visiting in Town": Gender Strategies in the Early Industrial West. *Historical Archaeology* 25(4):6–16.

Reeves, M. B., 1996, Autonomy and Status among African-Jamaican Slaves: An Examination of Jamaican Slave Settlements. Paper read at the 29th Annual Conference of the Society for Historical Archaeology, Cincinnati.

Reeves, M. B., 1997, *"By Their Own Labor": Enslaved Africans' Survival Strategies on Two Jamaican Plantations*. Ph.D. dissertation, Syracuse University, Syracuse.

Rice, P., 1994, The Kilns of Moquegua, Peru: Technology, Excavations, and Functions. *Journal of Field Archaeology* 21(2):325–344.

Rice, P., 1996a, The Archaeology of Wine: The Wine and Brandy Haciendas of Moquegua, Peru. *Journal of Field Archaeology* 23(2):187–204.

Rice, P., 1996b, Peru's Colonial Wine Industry and its European Background. *Antiquity* 70:785–800.

Rice, P., 1996c, Wine and Brandy Production in Late Colonial Moquegua, Peru: A Historical and Archaeological Investigation. *Journal of Interdisciplinary History* 27(3):455–479.

Rice, P., and Smith, G. C., 1989, The Spanish Colonial Wineries of Moquegua, Peru. *Historical Archaeology* 23(2):41–49.

Riordan, T. B., and Adams, W. H., 1985, Commodity Flows and National Market Access. *Historical Archaeology* 19(2):5–18.

Rothschild, N., 1990, *New York City Neighborhoods*. Academic Press, San Diego.

Roughley, T., 1823, *The Jamaica Planter's Guide*. Longman, Hurst, Rees, Orme and Brown, London.

Rowntree, L., and Conkey, M., 1980, Symbolism and the Cultural Landscape. *Annals of the American Association of Geographers* 70:459–474.

Rubertone, P. E., 1989, Landscape as Artifact: Comments on "The Archaeological Use of Landscape Treatement in Social, Economic and Ideological Analysis." *Historical Archaeology* 23(1):50–54.

Sanford, D., 1990, The Gardens at Germanna, Virginia. In *Earth Patterns: Essays in Landscape Archaeology*, edited by W. Kelso and R. Most, pp. 43–57. University Press of Virginia, Charlottesville.

Satchell, V., 1990, *From Plots to Plantations: Land Transactions in Jamaica, 1866–1900*. Institute of Social and Economic Research, Kingston.

Schama, S., 1989, *Citizens: A Chronicle of the French Revolution*. Knopf, New York.

Schrire, C., 1991, The Historical Archaeology of the Impact of Colonialism in Seventeenth Century South Africa. In *Historical Archaeology in Global Perspective*, edited by L. Falk, pp. 69–96. Smithsonian Institution Press, Washington, DC.

Schrire, C., 1992, Digging Archives at Oudepost I, Cape, South Africa. In *The Art and Mystery of Historical Archaeology*, edited by A. Yentsch and M. Beaudry, pp. 361–372. CRC Press, Boca Raton.

Schuyler, R. (Ed.), 1980, *Archaeological Perspectives on Ethnicity in America: Afro-American and Asian American Culture History*. Baywood, Farmingdale, NY.

Scott, E. (Ed.), 1994a, *Those of Little Note: Gender, Race, and Class in Historical Archaeology*. University of Arizona Press, Tucson.

Scott, E., 1994b, Through the Lens of Gender: Archaeology, Inequality, and Those "of Little Note". In *Those of Little Note: Gender, Race, and Class in Historical Archaeology*, edited by E. Scott, pp. 3–24. University of Arizona Press, Tucson.

Seifert, D. J., 1991, Within Sight of the White House: The Archaeology of Working Women. *Historical Archaeology* 25(4):82–108.

Shackel, P. A., 1993, *Personal Discipline and Material Culture: An Archaeology of Annapolis, Maryland, 1695–1870*. University of Tennessee Press, Knoxville.

Shackel, P. A., 1994, Town Plans and Everyday Material Culture: An Archaeology of Social Relations in Colonial Maryland's Capital Cities. In *Historical Archaeology of the Chesapeake*, edited by P. A. Shackel and B. J. Little, pp. 85–100. Smithsonian Institution Press, Washington, DC.

Shackel, P. A., 1996, *Culture Change and the New Technology: An Archaeology of the Early American Industrial Era*. Plenum Press, New York.

Shanks, M., and Tilley, C., 1987, *Social Theory and Archaeology*. Cambridge University Press, London.

Sheridan, R. B., 1974, *Sugar and Slavery: An Economic History of the British West Indies, 1623–1775*. Johns Hopkins University Press, Baltimore.

Sheridan, R. B., 1985, *Doctors and Slaves*. Cambridge University Press, London.

Singleton, T. (Ed.), 1985a, *The Archaeology of Slavery and Plantation Life*. Academic Press, Orlando.

Singleton, T., 1985b, Introduction. In *The Archaeology of Slave and Plantation Life*, edited by T. Singleton, pp. 1–14. Academic Press, San Diego.

Singleton, T., 1991, The Archaeology of Slave Life. In *Before Freedom Came: African-American Life in the Antebellum South*, edited by J. Edward and D. C. Campbell, pp. 155–191. University Press of Virginia, Charlottesville.

Soja, E., 1989, *Postmodern Geographies: The Reassertion of Space in Critical Social Theory*. Verso, London.

South, S. (Ed.), 1994, *Pioneers in Historical Archaeology: Breaking New Ground*. Plenum Press, New York.

Spencer-Wood, S. M., 1987, Miller's Indices and Consumer-Choice Profiles: Status-Related Behaviors and White Ceramics. In *Consumer Choice in Historical Archaeology*, edited by S. M. Spencer-Wood, pp. 321–358. Plenum Press, New York.

Spencer-Wood, S. M., and Heberling, S. D., 1987, Consumer Choices in White Ceramics: A Comparison of Eleven Early Nineteenth-Century Sites. In *Consumer Choice in Historical Archaeology*, edited by S. M. Spencer-Wood, pp. 55–84. Plenum Press, New York.

Staski, E. (Ed.), 1987, *Living in Cities: Current Research in Urban Archaeology, Special Publication Series Number 5*. Society for Historical Archaeology, Ann Arbor.

Stephen, J., 1969 [1824], *The Slavery of the British West India Colonies Delineated, as Exists Both in Law and Practice, and Compared with That of Other Countries, Ancient and Modern*. Kraus Reprint Co., New York.

Stewart, J., 1969 [1823], *A View of the Past and Present State of the Island of Jamaica*. Negro University Press, New York.

Stewart, J., 1971 [1808], *An Account of Jamaica and its Inhabitants*. Books for Libraries Press, Freeport.

Stinchcombe, A., 1995, *Sugar Island Slavery in the Age of Enlightenment*. Princeton University Press, Princeton.

Sweezy, P., 1942, *The Theory of Capitalist Development*. Monthly Review Press, New York.

Taylor, H., 1885, *Autobiography of Henry Taylor*. Longmans, Green, New York.

Thomas, D. H., 1993, The Archaeology of Mission Santa Catalina de Guale: Our First 15 Years. In *The Spanish Missions of La Florida*, edited by B. G. McEwan, pp. 1–34. University Press of Florida, Gainesville.

Thome, J. A., and Kimball, J. H., 1838, *Emancipation in the West Indies. A Six Months' Tour in Antigua, Barbadoes, and Jamaica in the Year 1837*. American Anti-Slavery Society, New York.

Thompson, V. B., 1987, *The Making of the African Diaspora in the Americas, 1441–1900*. Longman, New York.

Tilley, C., 1990, Michel Foucault: Towards an Archaeology of Archaeology. In *Reading Material Culture: Structuralism, Hermeneutics, and Post-Structuralism*, edited by C. Tilley, pp. 281–347. Basil Blackwell, Oxford.

Tomich, D., 1990, *Slavery in the Circuit of Sugar: Martinique and the World Economy, 1830–1848*. Johns Hopkins University Press, Baltimore.

Trigger, B. G., 1989, *A History of Archaeological Thought*. Cambridge University Press, London.

Trinkley, M., Adams, N., and Hacker, D., 1992, *Landscape and Garden Archaeology at Crowfield Plantation: A Preliminary Examination*. Chicora Foundation, Columbia.

Trouillot, M., 1982, Motion in the System: Coffee, Color, and Slavery in the Eighteenth-Century Saint-Domingue. *Review* 5:331–388.

Trouillot, M., 1993, Coffee Planters and Coffee Slaves in the Antilles: The Impact of a Secondary Crop. In *Cultivation and Culture: Labor and the Shaping of Slave Life in the Americas*, edited by I. Berlin and P. D. Morgan, pp. 73–98. University Press of Virginia, Charlottesville.

Upton, D., 1986, Vernacular Domestic Architecture in Eighteenth-Century Virginia. In *Common Places: Readings in American Vernacular Architecture*, edited by D. Upton and J. M. Vlach, pp. 313–335. University of Georgia Press, Athens.

Upton, D., 1988, White and Black Landscapes in Eighteenth-Century Virginia. In *Material Life in America, 1600–1860*, edited by R. B. St. George, pp. 357–369. Northeastern University Press, Boston.

Upton, D., 1992, The City as Material Culture. In *The Art and Mystery of Historical Archaeology*, edited by A. Yentsch and M. Beaudry, pp. 51–74. CRC Press, Boca Raton.

Wall, D. D., 1991, Sacred Dinners and Secular Teas: Constructing Domesticity in Mid-19th-Century New York. *Historical Archaeology* 25(4):69–91.

Wall, D. D., 1994, *The Archaeology of Gender: Separating the Spheres in Urban America*. Plenum Press, New York.

Wallerstein, I., 1974, *The Modern World System I: Capitalist Agriculture and the Origins of the European World-Economy in the Sixteenth Century*. Academic Press, New York.

Wallerstein, I., 1979, *The Capitalist World-Economy*. Cambridge University Press, London.

Wallerstein, I., 1980, *The Modern World System II: Mercantilism and the Consolidation of the European World Economy, 1600–1750*. Academic Press, New York.

Wallerstein, I., 1989, *The Modern World System III: The Second Era of Great Expansion of the Capitalist World-Economy, 1730–1840*. Academic Press, San Diego.

Walvin, J., 1980, The Rise of British Popular Sentiment for Abolition. In *Anti-Slavery, Religion and Reform*, edited by C. Bolt and S. Drescher, pp. 149–162. William Davidson and Sons, Kent.

Walvin, J., 1994, *Black Ivory: A History of Black Slavery*. Howard University Press, Washington, DC.

Watts, D., 1987, *The West Indies: Patterns of Development, Culture, and Environmental Change since 1492*. Cambridge University Press, London.

Werlen, B., 1993, *Society, Action and Space: An Alternative Human Geography*. Routledge, London.

Williams, E., 1944, *Capitalism and Slavery*. Russell & Russell, New York.

Williams, E., 1970, *From Columbus to Castro: The History of the Caribbean, 1492–1969*. Harper and Row, New York.

Wilmot, S., 1984, Not "Full Free": The Ex-Slaves and the Apprenticeship System in Jamaica, 1834–38. *Jamaica Journal* 17:2–10.

Wilson, M., 1929, Journal of a West Indian Proprietor: Introduction. In *Journal of a West Indian Proprietor, 1815–17, Edited with an Introduction by Mona Wilson*, pp. 1–13. Houghton Mifflin, Boston.

Wobst, H. M., 1977, Stylistic Behavior and information exchange. In *For the Director: Research Essays in Honor of James B. Griffin*, edited by C. E. Cleland, pp. 317–342. The University of Michigan Museum of Anthropology, Ann Arbor.

Wobst, H. M., 1978, The Archaeo-Ethnology of Hunter–Gatherers, or the Tyranny of the Ethnographic Record in Archaeology. *American Antiquity* 43:303–309.

Wobst, H. M., 1989, The Origination of Homos Sapiens, or the Invention, Control, and Manipulation of Modern Human Nature. Paper read at Wenner-Gren Symposium, No. 108: Critical Approaches in Archaeology: Material Life, Meaning and Power, Cascais, Portugal.

Wolf, E., 1982, *Europe and the People without History*. University of California Press, Berkeley.

Wright, G., 1995, *France in Modern Times: From the Enlightenment to the Present*, 5th edition. Norton, New York.
Wright, P. (Ed.), 1966, *Lady Nugent's Journal of Her Residence in Jamaica from 1801–05*. Institute of Jamaica, Kingston.
Wurst, L., 1991, "Employees Must Be of Moral and Temperate Habits": Rural and Urban Elite Ideologies. In *The Archaeology of Inequality*, edited by R. H. McGuire and R. Paynter, pp. 125–149. Basil Blackwell, Oxford.
Yamin, R., and Metheny, K. B. (Eds.), 1996, *Landscape Archaeology: Reading and Interpreting the American Historical Landscape*. University of Tennessee Press, Knoxville.
Yentsch, A., 1991, The Symbolic Divisions of Pottery: Sex-Related Attributes of English and Anglo-American Household Pots. In *The Archaeology of Inequality*, edited by R. H. McGuire and R. Paynter, pp. 192–230. Basil Blackwell, Oxford.

Index

"*f*" and "*t*" suffixes indicate that terms can be found in figures or tables on the pages indicated.